How to Build Speaker Enclosures

by
Alexis Badmaieff
& Don Davis

HOWARD W. SAMS & CO., INC.
THE BOBBS-MERRILL CO., INC.
INDIANAPOLIS • KANSAS CITY • NEW YORK

FIRST EDITION

EIGHTH PRINTING—1972

International Standard Book Number: 0-672-20520-3
Library of Congress Catalog Card Number: 66-29405

Preface

A loudspeaker without an enclosure is like a pump piston without its casing. In both, the casing provides a sort of "leverage" for the piston so that the piston (loudspeaker) can efficiently perform its work. Many very efficient "sound pistons," or speakers, are available today. Effective use of them in a manner which ensures that the most air will be pumped in an efficient and controllable fashion, however, depends almost entirely on the enclosure chosen.

The authors sincerely hope that this book will serve as a faithful guide to those seekers of realistic sound reproduction. The various chapters discuss the subject of speaker enclosures from the simple to the complex. Construction details are included for enclosures to fit practically any application and price range. In addition, the historical stages that acoustical engineering has passed through on the way to today's achievements are explained.

As is true in so many human endeavors, an honest effort to understand and utilize basic principles will be rewarded by more personally satisfying end results—in this case, a better performing music-reproducing system.

This book is dedicated to those who wish to know "why" as well as "how" and who have the inherent desire for perfection in all that they do, which is characteristic of the true audiophile.

ALEXIS BADMAIEFF
DON DAVIS

Contents

CHAPTER 5

CHAPTER 6

CHAPTER 7

CHAPTER 8

1

Speaker Enclosures

Designing, building, adjusting, and enjoying your own speaker enclosure can be an exciting and creative experience. Like most adventures into the world of "do-it-yourself," however, such a project can also be fraught with danger, such as selecting the wrong design, using faulty construction techniques, or failing to correctly adjust the system which you have built.

SPEAKER ENCLOSURES DEFINED

One of the most common fallacies about the design and construction of a speaker enclosure is that the enclosure is some sort of a musical instrument—that it should "resonate" or have "tone." Various types of woods are thought to be more suitable because "they are used in musical instruments"; or various shapes are chosen because "they are shaped like a timpani or a piano sounding board," etc.

The first clear distinction that must be made with regard to speaker enclosures, then, is that they are not musical instruments and that they do not generate musical tones. They are more accurately defined as *precision reproducers of musical tones.* This means that ideally they will add no tonal coloration of their own, but will remain passive, responsive only to the controlling input signal. The speaker and enclosure are as impersonal as a mirror. Any startling effects emanating from them must be only those produced by the original recording artist.

TWO FACTORS TO CONSIDER

In order to construct a speaker enclosure, two factors must be considered in conjunction with each other: (1) the speaker or *driver* and (2) the baffle or *house* in which the driver is mounted. The approach can be from either of two directions: (1) the speaker chosen will determine the type of enclosure to be constructed, or (2) the choice of enclosure desired will dictate the speaker required.

Where decor, size, and limited available placements in a room are the ruling factor, good design consists of selecting the enclosure that best answers the problem at hand and of carefully matching the optimum speaker to it (see Fig. 1-1).

If performance is the main objective, the speaker should be chosen first, and then an enclosure designed which will provide maximum utilization of the potentials inherent in the driver selected (see Fig. 1-2).

Today's designer is afforded a marvelous variety of acoustical devices from which to choose. Low-frequency speakers range from

Fig. 1-1. Example of small enclosures where compromise chosen favored room decor rather than performance.

Courtesy John Hilliard

Fig. 1-2. A house built specifically around a large speaker system where performance is the main objective.

8 inches to over 30 inches in diameter, with every degree of "hard" to "soft" suspension available in each size. The most commonly used sizes are 8 inches, 12 inches, and 15 inches. High-frequency drivers can be conic, ionic, and electrostatic or compression types.

FIVE BASIC ENCLOSURE TYPES

A speaker by itself causes interference with its own output. This is primarily due to the fact that the surface which moves the air (called a *diaphragm* or *cone*) causes the air to move both in front and in back of it. These simultaneous impulses generated by the motion of the cone are, unfortunately, "out-of-phase." This means that if they are allowed to meet, they will cancel each other. Therefore, if the front of the diaphragm/cone is not separated from the back of the diaphragm/cone, a large percentage of the energy fed into the speaker will be wasted in a fight between the air moved on either side of the diaphragm/cone surface.

There are five basic forms that this separation can take:

1. *Finite Baffle*—flat baffles, open-backed cabinets.
2. *Infinite baffle*—walls between rooms, large enclosures totally sealed.
3. *Bass reflex*—where the radiation from the rear of the cone is usefully added to the radiation off the front of the cone by means of phase inversion.
4. *Horn projectors.*
5. Combinations of any of the above.

Finite Baffle

Finite baffles are most commonly encountered in furniture-type console radios and consist of a board about 2 feet square, with the speaker mounted in the center. This forces the back radiation from the diaphragm/cone to travel a longer path to reach the other side. The larger the board used, the lower the *cutoff* frequency (frequency at which cancellation starts to occur). The wavelength of the cutoff frequency can be calculated as follows:

$$\lambda = \frac{V}{F}$$

where,

λ = wavelength in feet,
V = velocity of sound in air (about 1140 ft/sec),
F = frequency of sound in Hertz (Hz)*.

It can be seen that to prevent cancellation of bass frequencies at 30 Hz, the board would have to be about 38 feet square, or 19 feet each side of center.

Infinite Baffles

If the speaker were mounted on a board of infinite length and width, the back radiation would never meet the front radiation, and no cancellation could take place. Then the only determining factor would be the ability of the speaker to move enough air at very low frequencies to allow it to be audible.

An infinite baffle can be constructed by mounting a speaker in a wall so that the front of the diaphragm/cone is in an entirely different room. The area into which the back of the diaphragm/cone exhausts may be as small as 15 to 20 cubic feet if the surfaces are made nonreflective to sound waves with fuzzy soft material such

* Hertz (Hz) = cycles per second.

as glass wool, *Kimsul*, etc. (See Fig. 1-3 for a good example of a commercially available infinite baffle.)

Closet doors, when the closet has a substantial amount of clothes hung in it, can be used as a mounting surface for an infinite baffle. When the door is closed and gasketed, it becomes an infinite baffle. Of course, service presents no problem because of the easy access.

Courtesy R. T. Bozak Mfg. Co.

Fig. 1-3. These large enclosures (4 ft x 3 ft x 1½ ft) represent the smallest an infinite baffle should be made for good efficiency.

A special technique used in many of today's smaller speaker systems compresses the infinite baffle idea into a 2-cubic-foot box (see Fig. 1-4). In this case, the air that the back of the cone compresses, also serves as an additional "spring" in conjunction with the one mechanically designed into most speakers. The drawback to this method is very low efficiency as a result of having to substantially increase the cone's mass as part of the design process.

Speaker enclosures, in conformance with other facts of life, simply don't allow "something for nothing."

Bass-Reflex Enclosures

Another approach is to add the radiation from the rear of the diaphragm/cone to the radiation from the front. This is done by using the volume of air in the enclosure which acts in conjunction with the mass of air entrapped in a tuned port hole to create an in-phase, additive relationship.

This combination of the rear radiation being added in phase to the radiation off the front of the cone results in almost twice the output for a given excursion of the cone than would be expected if the speaker were mounted in an infinite baffle.

Courtesy Altec Lansing, Div. LTV Ling Altec, Inc.

Fig. 1-4. An infinite baffle reduced to a very finite size (2 cu ft) by the use of a soft-suspension woofer.

These phase-inversion or bass-reflex enclosures offer greatly increased bass response with a minimum of structural and tuning work (see Fig. 1-5).

Horn Projectors

Bass response can be substantially increased if the front of the cone is coupled to a long, expanding horn. The old Edison phonographs had a diaphragm about the size of a dime. When the large "morning glory" horn was attached to the small diaphragm, each motion of the dime-sized surface was transformed from a small-area radiation into a large-area radiation. This means that horns act as acoustical transformers. Such transformers can take many shapes such as conical, hyperbolic, catenary, parabolic, and exponential (see Fig. 1-6).

In all types of horns, the effect is to present to the small diaphragm at the throat a very high but consistent acoustical impedance while transforming the high-pressure, low particle-velocity wave from the surface of the cone into a low-pressure, high particle-velocity

Fig. 1-5. Optimum size (7 cu ft) bass-reflex enclosure.

wave with a low impedance, matching that of the air in the room as it reaches the mouth of the horn.

This consistent high impedance at the throat of the horn can be maintained as long as twice the square root of the mouth area times pi (π) is greater than the wavelength required. In other words, when $2\sqrt{\pi MA}$ is greater than λ, it is a useful horn. If $2\sqrt{\pi MA}$ is equal to or less than λ, then the horn does not properly "load" the diaphragm at the throat to the air impedance at the mouth.

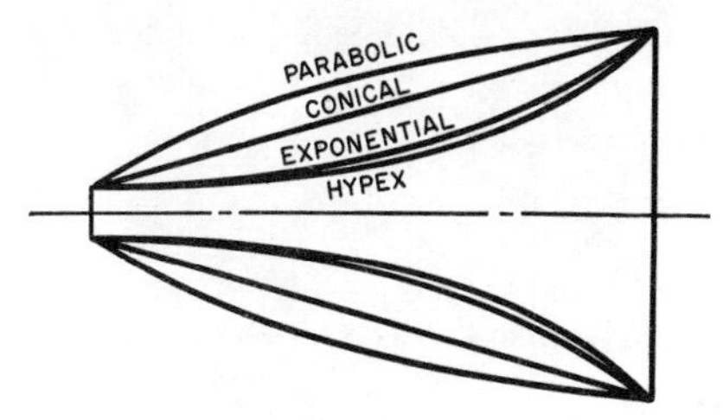

Fig. 1-6. Comparison of the different rates of flare used in the design of horn projectors.

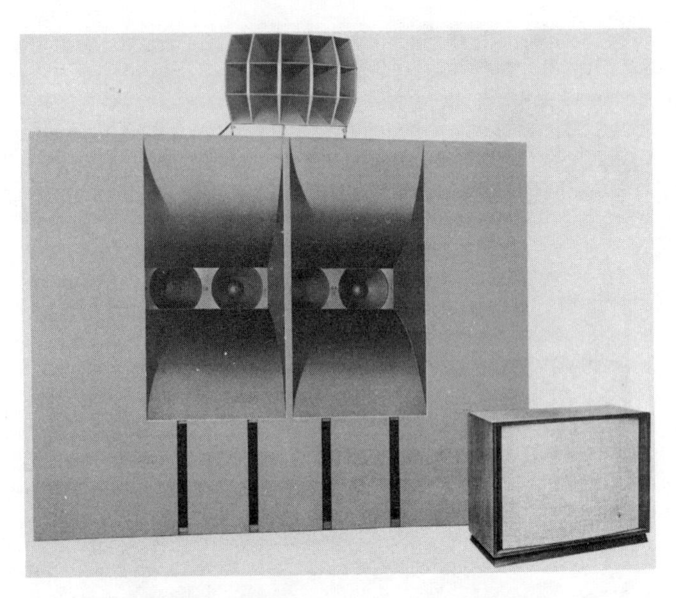

Fig. 1-7. Overwhelming size of a large, straight horn.

From this it can be seen that to have a usable low-frequency horn, large dimensions are required. For 30 Hz, a mouth area of 45 square feet would be needed. Nine times five feet is a large mouth in any living room (see Fig. 1-7).

Courtesy Klipsch and Associates, Inc.

Fig. 1-8. A folded corner horn which uses the natural flare of the room corner as its terminating mouth.

Folded corner horns can materially reduce the enclosure dimensions required to achieve such a large mouth area by folding the horn back and forth in its own path before using the corner of the room as the mouth of the horn (see Fig. 1-8).

A horn device at frequencies above its low cutoff point, or f_c, gives very effective control of the polar pattern or directionality.

For all these reasons, a horn-loaded speaker mechanism is very efficient; i.e., it will provide more acoustical watts output per electrical watts input.

Combinations

Combinations of these basic forms make up most of the commercial offering in the high-fidelity market today. Economy, decor, and associated equipment limit ideal theoretical considerations.

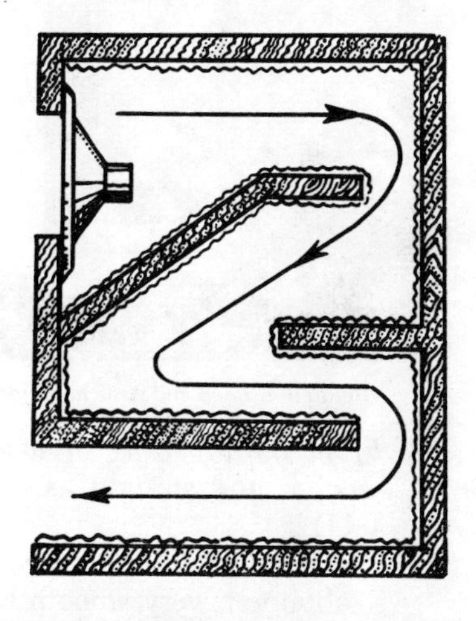

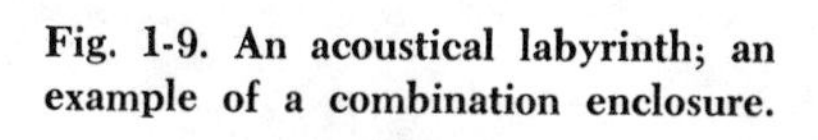

Fig. 1-9. An acoustical labyrinth; an example of a combination enclosure.

The result is an engineered system whose compromises are not detectable under actual home listening conditions, even in direct comparison with so-called theoretical perfection. This is particularly true because no speaker system has yet achieved anything close to perfection, and while some come closer than others, it is still a question of how much distortion your personal ear enjoys or rejects the most. One person may be tone deaf but appreciate dynamic range. Another may be quite sensitive to pitch (which would require an expensive turntable) and yet completely ignore distortion, compression of dynamic range and regularity of frequency response.

Various popular combinations include:

1. *Acoustical labyrinth* (see Fig. 1-9).

Advantages: Increased bass loading over extended range.
Disadvantages: Large size; out of phase at crossover.

2. *Back-loaded horn.* This is really a bass-reflex device in which a horn has been placed on the port (see Fig. 1-10).
   Advantages: Good, efficient bass response.
   Disadvantages: Rough response; phasing at crossover difficult.

Fig. 1-10. A back-loaded corner horn.

Courtesy Klipsch and Associates, Inc.

3. *Front horn-loading of woofer*, bass-reflex loading of back of woofer, and straight axis horn for high frequencies (see Fig. 1-11).
   Advantages: Proper phasing can be achieved; high efficiency obtained; very smooth response; exceptional polar response.
   Disadvantages: Unless size is quite large, bass response falls off below 40 Hz; transient response slightly less than that of a horn woofer.

These samples of commercially obtainable systems illustrate that almost any conceivable combination could be tried, and a quick look at the back issues of the early *High Fidelity* magazine will show that indeed they have been.

DESIGN CONSIDERATIONS

This chapter, then, has discussed the important design considerations that should be taken into account before one decides on which of the five enclosure types to plan around. These considerations are:

Fig. 1-11. A horn and bass-reflex combination with straight, exponential, high-frequency horn mounted in a bass-reflex port area.

1. Decor and physical size
2. Efficiency
3. Smoothness of tonal response
4. Range of tonal response
5. Distortion in its many forms:
   A. Harmonic
   B. Intermodulation or Doppler
   C. Transient
   D. Phase
6. Polar characteristics

SUMMARY

Decor and physical size will be determined by use. The owner of a small apartment may desire to reproduce only string quartets (where the small infinite baffle with a "long throw" woofer is a good compromise); the owner of a 60-foot living room may wish the Boston Symphony present at a distance equal to first row center (where a large, straight, woofer horn carefully phased at the crossover to a multicellular or sectoral horn would create a hard-to-disbelieve illusion). In other words, those who listen to small folk-singing groups, string quartets, modern jazz combos, etc., can accept a degree of power inefficiency in their speakers, whereas those who listen to pipe organs, symphony orchestras, massed choirs, etc., cannot.

No one accepts roughness of tonal response except as a compromise to expense; and in the systems described in this book, every effort is made to achieve the smoothest tonal response possible at any size, price, or performance level.

Deliberate limitation of tonal range, however, is often used and can provide many benefits. In a small speaker, extended low-frequency response may be at the expense of excessive distortion, since the small woofer moves nonlinearly at the low frequencies. Much better sound can be achieved by limiting the speaker's low-frequency response at the frequencies where distortion starts to increase too rapidly. This can often be achieved by inexpensive high-pass filters either in the amplifier or even in the crossover network.

If both low distortion and wide dynamic range are desired, "think big." If low distortion is desired and dynamic-range compression can be tolerated, then "think small." Cost and decor limitations usually result in a compromise between distortion, efficiency, and less control of irregularities. The better the balance, the better is the combination. Much latitude is offered to the designer in the choice of such balances.

Polar response will be primarily a function of the high-frequency driver and horn chosen—and "horn" is the correct word if all persons in a listening area are to hear approximately the same quality of reproduction.

Proper phasing of the low- and high-frequency units is a little-known art, and it will be discussed in detail in the chapter on infinite baffles and also in the chapter on crossover networks. Suffice it to say that without very correct phasing of high- and low-frequency drivers, transients will never be properly reproduced. Sound travels at about 1140 ft/sec, and the ear can detect as little as three milliseconds difference in short duration sounds (castanets for example). Since sound travels 1.14 feet per millisecond, sounds having less than a three-foot path difference can be heard.

While the variety of problems to consider and the infinite number of possible answers can at first seem confusing, approaching each basic design one at a time will allow today's knowledge of the art to fall into place naturally.

2

Drivers for Enclosures

Every enclosure alters in some way the performance of the driver placed in it. For best results it is therefore necessary to know the basic limitations of the driver.

BASIC INFORMATION

Table 2-1 illustrates some vital statistics that apply universally to cone-type woofer speakers. *Rated cone diameter* is the advertised size of the speaker. The three most common sizes are shown. While woofers as small as 3 inches and as large as 30 inches have been employed, good design criteria dictate one of the three sizes shown. If a woofer is smaller than 8 inches, it requires so many cones of that size to generate a usable *acoustical* power that phasing problems are encountered. This is due to their physical separation on the baffle board.

If the woofer is much larger than 15 inches, the mass of the moving parts becomes too great to retain effective control; and, therefore, poor transient response results.

Column B of Table 2-1 is *actual cone diameter*. This is the figure to be used in calculating the cone area of a standard size speaker:

$$A = \frac{\pi D^2}{4} - 8$$

where,

A = actual cone area,
π = 3.1416,
D = diameter of the circle.

Column C of Table 2-1 shows the effective piston areas. Column D shows the amount of peak-to-peak excursion (the distance from the farthest forward movement of the cone to its most rearward position) that would be demanded of any size of cone at 50 Hz if it were required to generate one acoustical watt. This figure has no fixed relation to electrical power required (which may vary from 2 watts to 500 watts) in order to drive the cone to such an excursion, but it is the *acoustical* power generated by the mechanical action of the piston moved that distance at that rate. It should be noted

Table 2-1. Basic Mounting and Performance Data for Cone Speakers

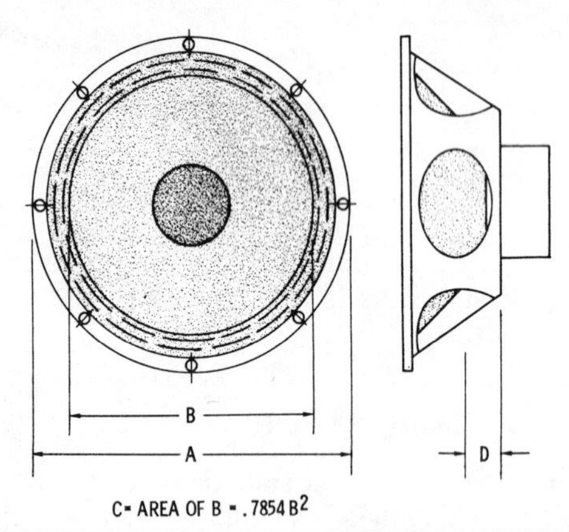

A Rated Cone Diameter	B Actual Cone Diameter	C Effective Piston Area	D Peak-to-Peak Cone Excursion Required to Generate One Acoustical Watt at 50 Hz*	E Volume of Displacement Inside of Enclosure
8 in	6.75 in	28 sq in	1.80 in	0.15 cu ft
12 in	10.50 in	78 sq in	0.80 in	0.40 cu ft
15 in	13.25 in	133 sq in	0.53 in	0.75 cu ft

* 0.50 inch is maximum allowable peak-to-peak excursion for acceptable efficiency and low distortion.

that 0.50 inch is the maximum excursion that can be considered without excessive loss of electrical-to-acoustical conversion efficiency.

The final column in Table 2-1 shows how many cubic feet must be subtracted from the calculated interior volume of the enclosure due to the physical volume displaced by the driver itself.

CHOOSING THE NUMBER OF DRIVERS AND THEIR SIZES

The following relations should be noted when one decides whether to use a larger or smaller driver versus a larger or smaller *number* of drivers:

A = cone area,
E = cone excursion,
F = lowest frequency desired,
P = acoustical power desired.

1. If the cone area is doubled, the cone excursion is halved:

$$2A = \frac{E}{2}$$

2. If the cone area is halved, the cone excursion is doubled:

$$\frac{A}{2} = 2E$$

3. If the frequency is halved, the cone excursion is increased by a factor of four:

$$\frac{F}{2} = 4E$$

4. If the frequency is doubled, the cone excursion is reduced by a factor of four:

$$2F = \frac{E}{4}$$

5. If the cone excursion is halved, the acoustical power is halved:

$$\frac{E}{2} = \frac{P}{2}$$

6. Conversely, if the excursion is doubled, the acoustical power is doubled:

$$2E = 2P$$

7. If the cone area is doubled and the cone excursion remains the same, the acoustical power increases by a factor of four:

$$AE = P$$
$$2AE = 4P$$

8. By the same token, if the cone area is halved and the cone excursion remains the same, the acoustical power decreases by a factor of four:

$$\left(\frac{A}{2}\right)E = \frac{P}{4}$$

Using these parameters one can determine the number of drivers desired to ensure achievement of at least one acoustical watt at 50 Hz with low distortion and a peak-to-peak cone excursion less than 0.50 inch:

Number of Drivers Required	*Size of Drivers*
5	8 inches
3	12 inches
2	15 inches

The maximum acoustical power each size can produce at 50 Hz without exceeding 0.50 inch peak-to-peak cone excursion is as follows:

Size of Driver	*Acoustical Power*
8 inches	0.06 watt
12 inches	0.40 watt
15 inches	0.95 watt

To mount multiple drivers as woofers, care should be taken to ensure that they are close together. They should operate as nearly as possible as a single diaphragm. The high-frequency drivers should also be mounted close to the low-frequency woofers to minimize phase differences that can occur at the listener's ear if the angle between high frequencies and low frequencies becomes too large.

EFFICIENCY OF THE DRIVER

Changes in cone mass affect efficiency above the point of cone resonance. Tripling the cone mass reduces the acoustical power output by one-half. Conversely, reducing the cone mass by one-third would increase the acoustical power output by a factor of two to one, or 3 db.

While reduction of cone mass increases efficiency it also increases distortion. Cone mass is therefore chosen as a compromise between efficiency and distortion requirements. Magnet weight, or magnet mass, changes the efficiency of the driver. Magnet flux is directly proportional to magnet weight and/or volume. The efficiency of a speaker is directly proportional to the flux up to a limit of around 20,000 gausses where saturation of the magnetic circuit occurs. This means, if all other factors remained constant, that doubling the magnet size would approximately double the efficiency of the driver.

It is very possible in actually comparing different drivers to find, for example, that Speaker A has a magnet twice the size of Speaker B, and yet Speaker A has an efficiency one-half that of Speaker B. This could be because Speaker A may have a much greater cone mass, its voice-coil gap may be much larger, its voice-coil resistance may be higher, or it may have a mechanical suspension that is stiffer and more difficult to overcome.

All parameters of the driver—cone area, cone excursion, weight of moving system, magnet size, voice-coil material, size and gap, and the form of mechanical suspension employed—are interwoven with the enclosure design to obtain a speaker system.

MATCHING DRIVER AND ENCLOSURE

The designer of a successful speaker enclosure cannot change the characteristics of the driver chosen (unless he is also a speaker manufacturer), but must rely on the action of the driver in the enclosure to adjust those parameters needing change.

In using the information gathered here to select a suitable woofer-driver for an enclosure, the following material must be calculated or obtained from the manufacturer:

1. How great an excursion of the cone, for the size driver being contemplated, would be required at the lowest frequency and highest sound pressure level (spl) desired?
2. At the lowest frequency and highest spl desired, is the electro-acoustical efficiency of the driver high enough to allow:
   A. A commercially available amplifier to be used?
   B. The required sound pressure to be reached without exceeding the power capabilities of the speaker?
3. If a smaller diameter driver is chosen, does the number of drivers required for sufficient bass response lead to detrimental phase relations as the frequency is increased?
4. Does the driver chosen retain linear response, low distortion, and acceptable polar characteristics at least to the nominal crossover frequency?

In considering all these factors, realize that no single unit can possess all the qualities required. Engineering design consists of balancing the choices available to achieve the greatest harmony between conflicting requirements.

A LOW-COST EXAMPLE

An excellent example of a low-cost, small-size, wide-range 8-inch speaker is shown in Fig. 2-1. This speaker has a large magnet to

Fig. 2-1. A well designed, all-range 8-inch speaker.

Courtesy Gil Evans

Fig. 2-2. A system built around two of the drivers shown in Fig. 2-1.

help raise efficiency, a large cone excursion to allow useful amounts of air to be moved at low frequencies, a subdivided cone to allow minimum variations in amplitude response as frequency is increased, and a flat cone in order to avoid a restriction in the angle of coverage at high frequencies. In fact, it manifests the perfect balance of available choices in a single 8-inch driver. This driver was originally the design project of the Bell Telephone Laboratories when they first investigated dynamic speakers in the 1930's, and through constant redesign it has continued to be the best speaker of its size. Fig. 2-2 shows how this simple speaker in an infinite baffle is used in an elaborate system.

IMPORTANCE OF BALANCE

An important factor to consider in speakers is balance. Regardless of whether or not the full range or less than the full range is to be reproduced, the multiplication of the highest frequency by the lowest frequency to be reproduced should equal a figure close to 500,000. For example, a really full-range unit operating from 25 to 20,000 Hz equals 500,000. A system that operates from 35 to 15,000 Hz equals 525,000. If the highest frequency to be reproduced were chosen as 10,000 Hz, then 50 Hz would be preferred as the lowest frequency, and 60 Hz the highest compromise at the low-frequency end of the system's response.

MULTIPLE ARRAYS

In order to produce the same acoustical power at 50 Hz, a 15-inch speaker cone will have to move only about one-half the distance that a 12-inch cone has to move. Two 12-inch cones mounted closely together can produce the same acoustical power with one-half the cone excursion for each unit.

When speakers are used in multiples for the same frequency range, danger of phasing problems arise. Phase should not be confused with polarity. It is a simple matter to ensure that four 12-inch speakers on a panel have the same polarity (that they all move forward at the same time and all move backward at the same time), but the phase difference of the radiation from the far left cone on the panel compared to the radiation from the far right cone on the panel can be substantial at certain frequencies and at certain listening positions in the room. To minimize these phasing problems, multiple-cone arrays are arranged as shown in Fig. 2-4.

If two speakers are separated by an excessive distance, phase cancellation will take place when the listener is positioned at unequal distances from the two speakers (angular displacement). This

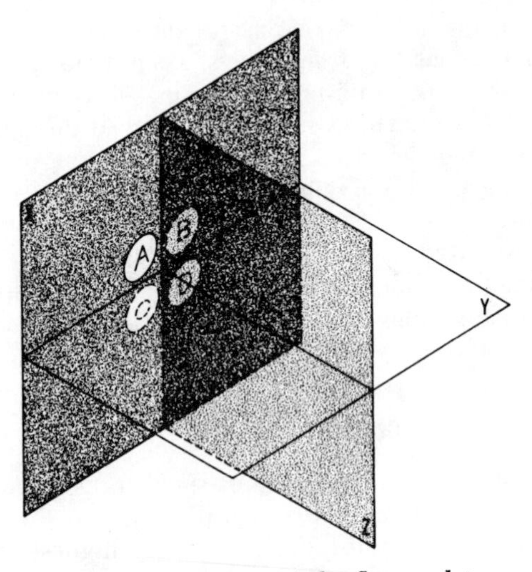

Fig. 2-3. The mirror images of the floor and two walls in a trihedral corner increase the bass efficiency of a single cone woofer.

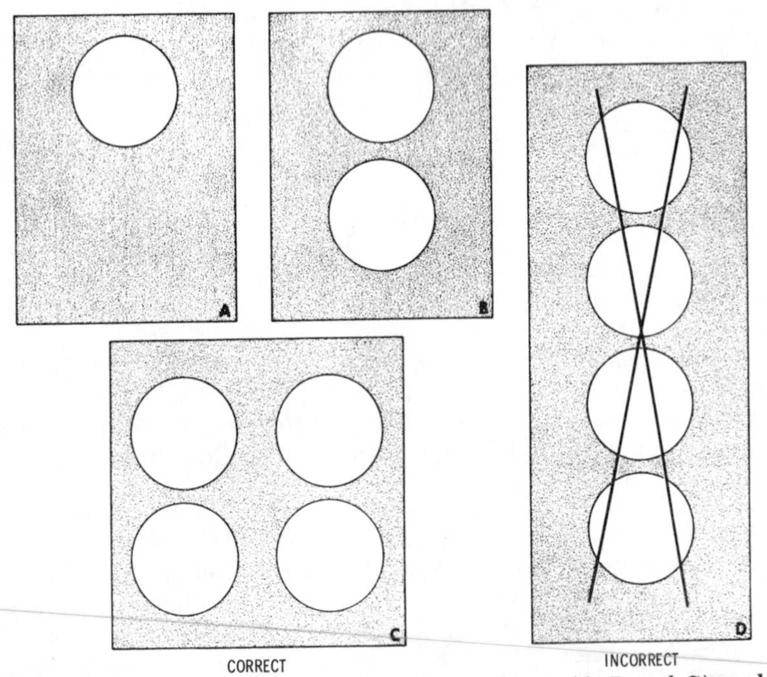

Fig. 2-4. Correct mounting configurations of cone drivers (A, B, and C), and an undesirable choice (D).

phase cancellation will occur at those frequencies where the difference in distance from one speaker to the listener is one-half wavelength of the distance from the listener to the second speaker.

It is this problem that rules out the use of many small but inexpensive speakers to obtain a high-performance speaker.

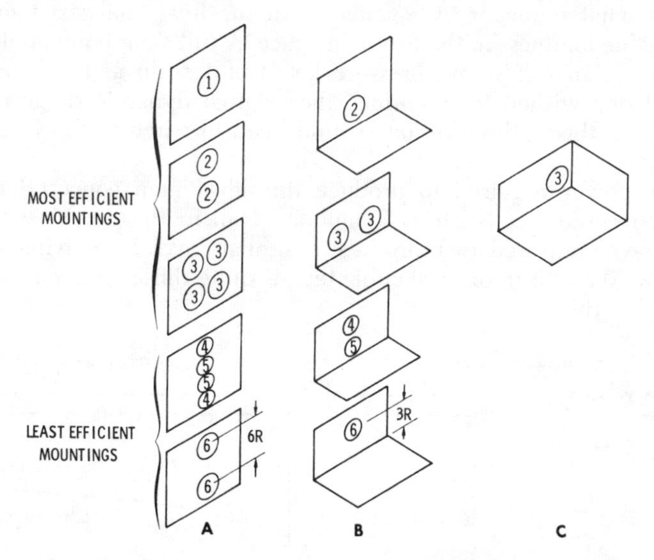

Fig. 2-5. Effect of midwall mounting (A), dihedral (floor and wall) mounting (B), and trihedral (two walls and the floor) mounting (C). Arrangements in adjacent columns produce the same sound pressure level.

PHASING WOOFER AND TWEETER

Still another aspect of phase relations has to be considered—time delays between drivers operating at the same frequency in the crossover region. Sound travels approximately 1140 feet per second. This means about 1.14 feet per millisecond—a foot every 1/1000 of a second. It has been demontrated that the ear can detect differences as short as three milliseconds on sounds such as castanets. This was strikingly illustrated when, in 1935, the movie sound track of the tap dancing of Eleanor Powell reproduced the taps with an added echo on the two-way speaker systems then in use. It was soon discovered that the low-frequency driver and the high-frequency driver used for these recordings were some eight feet apart due to the difference in length of the horns employed.

When the high-frequency unit was moved back to the point where both drivers were in the same vertical plane, the echo disappeared. Subsequent study of the problem proved that a delay of less than three milliseconds was not detectable in systems using crossovers in the region of 350 to 800 Hz (see Fig. 2-6).

EFFICIENCY

Just what is meant by *efficiency*? In the final analysis it means the usable loudness in the listening space. A full symphony orchestra can reach an spl (sound pressure level) of 120 db at the listener's ear. If one wished to reproduce the original dynamic range of an orchestra, this is the level one would need to reach in the listening room.

The power required to produce this level in a concert hall of 600,000 cubic feet (such as Symphony Hall in Boston, Mass.) and the power required to reproduce a similar level in a living room 30 ft × 20 ft × 8 ft, or 4800 cubic feet, is quite different even though the spl is the same.

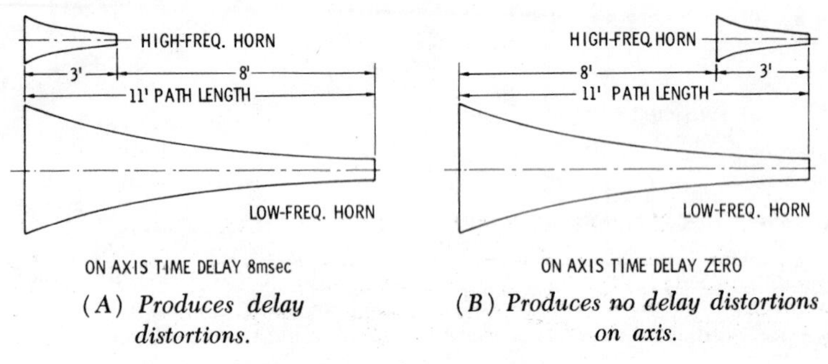

(A) *Produces delay distortions.*

(B) *Produces no delay distortions on axis.*

Fig. 2-6. Speaker arrangements.

Assuming that one is willing to sit as close as eight feet from the speaker system, then the goal would be 120-db spl peak intensity at eight feet from the system as a maximum acoustical power output. At eight feet, 0.4 acoustical watt would be approximately 100 db; 120 db would be 40 acoustical watts.

The large, built-in system described in Chapter 3, on infinite baffles, has a measured peak efficiency, in the 50- to 400-Hz range, of 40 percent (*plane-wave tube measurements*). This means that an amplifier peak power of 100 electrical watts would give an acoustical peak output of 40 watts from the speaker. It should be noted that most of the sheer power present in a musical passage that would reach 120-db spl would be concentrated well below 100 Hz, and as the ear does not hear sounds in the bass regions as efficiently as it does in the midrange, the loudness of these sounds is more "felt" than heard; but the power required is high. Yet there are on the market today, speaker systems with efficiencies as low as 0.1

percent. This means that for the same loudness at any given frequency, the less efficient system would require 400 times more power than the 40-percent efficient system.

In order to achieve 120-db spl at eight feet, with a speaker of 0.1-percent efficiency, the peak electrical input power would have to be 40,000 watts; if the speaker were 1-percent efficient, then 4000 watts would suffice; and if 10-percent efficient, 400 watts would do the job.

One also has to bear in mind that even if amplifiers were free, and one had 40,000 watts available, the speaker that would require such an input never is capable of handling such power. It usually is rated at a maximum of 30 watts before burnout. If enough units are put together to withstand power-handling problems, then the same type speaker no longer exists, but instead, a new multiple-driver type comes into being with all its attendant phase problems.

EFFICIENCY AND EQUALIZATION

One of the major disadvantages of low efficiency that often is not readily detected by the novice is the limitation it places on any chance to use a bass-boost control on the amplifier (often desirable because of a poor listening room and/or phonograph disc).

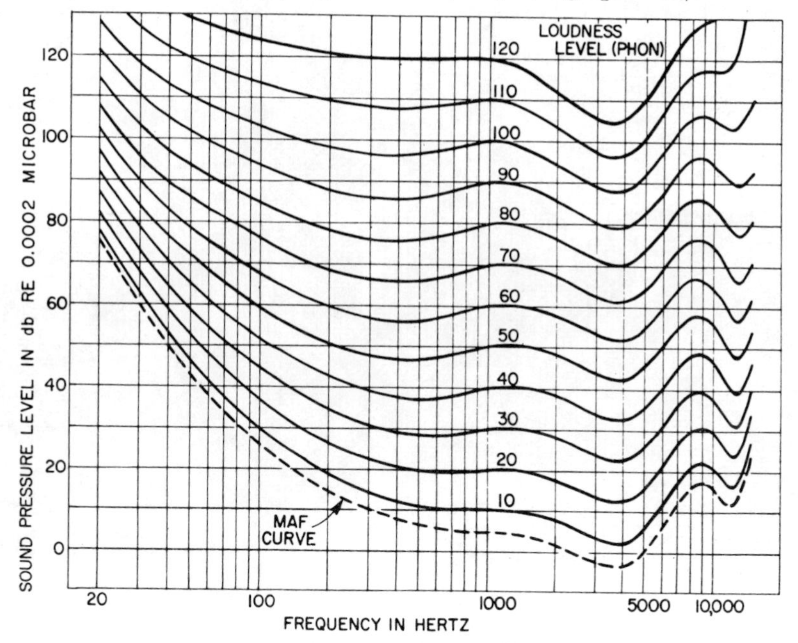

Fig. 2-7. Equal-loudness curves.

Consider a system that is 1-percent efficient. To produce 100-db spl at eight feet would require forty electrical watts from the amplifier (100-db spl at eight feet requires 0.4 acoustical watt from the speaker). Suppose still further that a 60-watt amplifier is being used—real continuous watts, not music power, peak, or other short-term values. In quality amplifiers, a 12-db bass-boost capability is usually considered conservative. A mere 6-db boost in bass response requires 160 electrical watts (40 watts times four), 100 watts more than the 60-watt amplifier can provide; 12-db boost would require 640 electrical watts. This means that the bass-boost controls are not really usable with the speaker selected.

Using a system that is 40-percent efficient would require one electrical watt for 100-db spl at eight feet; six db of bass boost would require four electrical watts; and if all 12-db bass boost available were used, the amplifier would be called on to produce 16 electrical watts. In this case the choice of speaker would allow full use of the capabilities of the amplifier chosen to power it.

If one either sits closer than eight feet (at four feet a six-db increase occurs) or accepts some compression of dynamic range,

Courtesy Altec Lansing, Div. LTV Ling Altec, Inc.

Fig. 2-8. Coaxial speaker incorporating a multicellular high-frequency driver horn.

efficiency becomes an allowable parameter to compromise. For example, the listener who enjoys folk music and small jazz combos finds that such groups seldom, if ever, exceed 100-db spl. Let's imagine that this same listener lives in a quiet apartment. The ambient noise in the listener's apartment is about 45-db spl total reading. This means that with a top level of 100-db spl there is 55 db

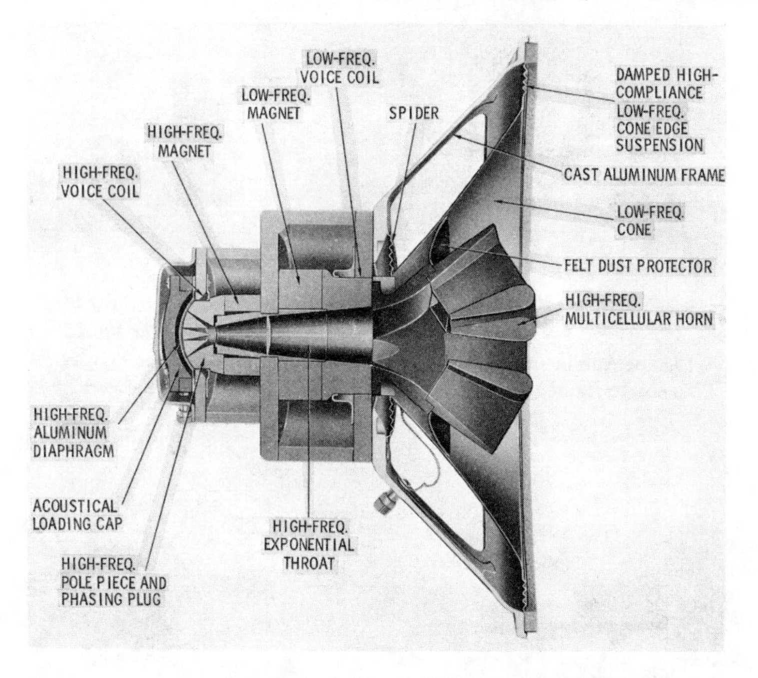

Couresy Altec Lansing, Div. LTV Ling Altec, Inc.

Fig. 2-9. Cutaway view of the coaxial speaker in Fig. 2-8 showing the phasing plug employed to assure equal distance from the high-frequency diaphragm to the throat of the multicellular horn.

of dynamic range available. This exceeds the dynamic range available on records and all but the most professional tape machines. (All recording processes compress the original in some manner.) To achieve this degree of dynamic range, the 1.0-percent efficient speaker needs a 40-watt amplifier; the 10-percent efficient speaker needs four watts; and the 40-percent efficient speaker will need only a 1-watt amplifier.

The choice of a maximum level of 100-db spl permits the listener to consider a speaker with efficiency as low as 1 percent.

Unfortunately, little information is available on speaker efficiency. That which is available specifies one-watt electrical input to produce

(*A*) *Speaker.* (*B*) *Matching network.*

Courtesy Jensen Mfg. Div., The Muter Company

Fig. 2-10. Speaker incorporating woofer, midrange, and tweeter drivers all on a single frame, acoustically balanced with a matching network.

Courtesy Jensen Mfg. Div., The Muter Company

Fig. 2-11. Cutaway view of a typical triaxial speaker.

Fig. 2-12. A heavy-duty woofer made especially for low frequencies.

x number of db spl at a given distance. For example, the 8-inch speaker shown in Fig. 2-1 has an efficiency specification of 95.5-db spl, at a distance of four feet, from one watt of electrical input. Since this speaker has a power-handling capability of 15 watts, the greatest level it can be expected to produce from 15 watts will be 107.2 db at four feet. The woofers used in the large built-in infinite-baffle system shown in Fig. 3-1 in Chapter 3, have a specification that reads: 103 db from one watt at four feet, or 118.5 db from a rated 35 watts at four feet. Since two speakers are used, 70 watts of power would yield 121.5-db spl at four feet.

(A) 500 Hz, with step high-frequency attenuation control.

(B) 800 Hz, with step attenuation control.

Courtesy Altec Lansing, Div. LTV Ling Altec, Inc.

Fig. 2-13. Crossover networks.

In short, be sure to give consideration to the anticipated power requirements of the system at equalized settings as well as at "flat" settings on the tone controls.

At this point still another factor enters the picture. It is a simple matter to get even a three-inch speaker to move back and forth within its limits at 30 Hz, but the amount of air which it moves remains totally inaudible. Fig. 2-7 shows that for a sound that is judged to be a given loudness and measures 60-db spl at 1000 Hz, a 30-Hz tone would have to be 90-db spl to be judged equally loud. At minimum audible frequencies (the quietest tones the ear can detect at a particular frequency) the softest tone the ear hears at 1000 Hz is about 5-db spl. However, a tone that is just audible at 30 Hz must be at least 60-db spl.

HIGH-FREQUENCY DRIVERS

High-frequency drivers are not within the scope of this book on enclosures because they do not normally require special housing; however, a few factors that relate to matching a suitable high-frequency unit to the woofer-enclosure combination chosen are:

Fig. 2-14. A typical sectoral horn.

Courtesy Altec Lansing, Div. LTV Ling Altec, Inc.

Fig. 2-15. A typical high-frequency driver.

1. The efficiency of the high-frequency unit should be close to that of the low-frequency unit so that the final exact match at crossover can be accomplished with a minimum or even a lack of attenuators.
2. The two units should be mounted so as to minimize differences in phase (distance from woofer diaphragm to the ear as compared to distance from high-frequency diaphragm to the ear).
3. The quality of tonal response, the polar pattern, and the impedance should harmonize with the low-frequency driver.

Coaxial and triaxial speakers, such as those illustrated in Figs. 2-8 through 2-11, solve all these problems inasmuch as they are mechanically, electrically, and acoustically integrated systems adjusted by the manufacturer.

If separate drivers are used, such as the woofer shown in Fig. 2-12, the networks shown in Fig. 2-13, and the high-frequency horn and driver shown in Figs. 2-14 and 2-15, then care in following the above will result in a well-engineered system.

If one is building his first enclosure, it is possible to greatly minimize the potential problems by choosing a coaxial or triaxial driver unit. For those who are ready to attempt simple measurements and are willing to do some experimenting on their own, the separate driver units offer a much wider latitude of performance.

3

Infinite Baffles

A number of years ago a friend of ours visited an avid high-fidelity fan and heard a confusing number of the then popular baffle names dropped in the course of the conversation. He later remarked that he had found the evening "infinitely baffling," which opened a discussion about the simplest type of baffle—the *infinite baffle.*

DEFINITION

While the purist views "infinite" with a proper sense of awe, let us corrupt our definition to refer to any enclosure that prevents the back radiation of the speaker cone from meeting the front radiation in a detrimental manner. By this definition, a large, flat board 50 ft × 50 ft qualifies if the speaker is mounted in its center; or it may be reduced to a box two cubic feet in volume if totally enclosed.

AN INFINITE BAFFLE ILLUSTRATED: SYSTEM 1

Let us consider the design requirements of the first system to be studied. In this system, performance was given first consideration, and as a result the entire living room in the home was specifically designed to house and acoustically enhance the musical system. Approximately 300 cubic feet were allowed for each 25-ton enclosure.

The enclosure dimensions are 4 ft × 4 ft × 18 ft. Three of the walls of the enclosure are solid concrete. The fourth or front wall of the enclosure is of 1-inch plywood reinforced with random diagonal bracing. Obviously the enclosure is meant to house a musical *reproduction* system and not to contribute tones of its own manufacture (see Fig. 3-1).

Courtesy John Hilliard

Fig. 3-1. Integrated planning produces an excellent combination of architecture and acoustics.

The choice of drivers, their placement on the baffle, and their relations to each other were arrived at after careful review of desired criteria relating to linearity, distortion, phase, efficiency, and polar response (see Figs. 3-2 and 3-3).

Table 2-1 in Chapter 2, Column C, shows the relation of speaker size to effective cone area. A pair of 15-inch woofers were chosen because:

1. They have a piston area of 266 square inches, which was suf-

ficient to meet the efficiency goal shown in Table 2-1, Column D.
2. They were small enough to have sufficiently low mass which ensured good transient response.
3. They were large enough to require only two units in order to

Courtesy John Hilliard

Fig. 3-2. Front view of a large infinite-baffle system using two woofers and a 500-Hz sectoral horn.

achieve adequate low-frequency output without excessive cone excursion.
4. Their response extended above 500 Hz, the desired crossover point for the high-frequency driver and horn chosen.

If an 18-inch or 30-inch woofer had been chosen, the necessary cone area would have been gained at the expense of response to transient sounds.

If four 12-inch drivers had been chosen, excellent transient response could have been achieved, but polar response would have been less predictable.

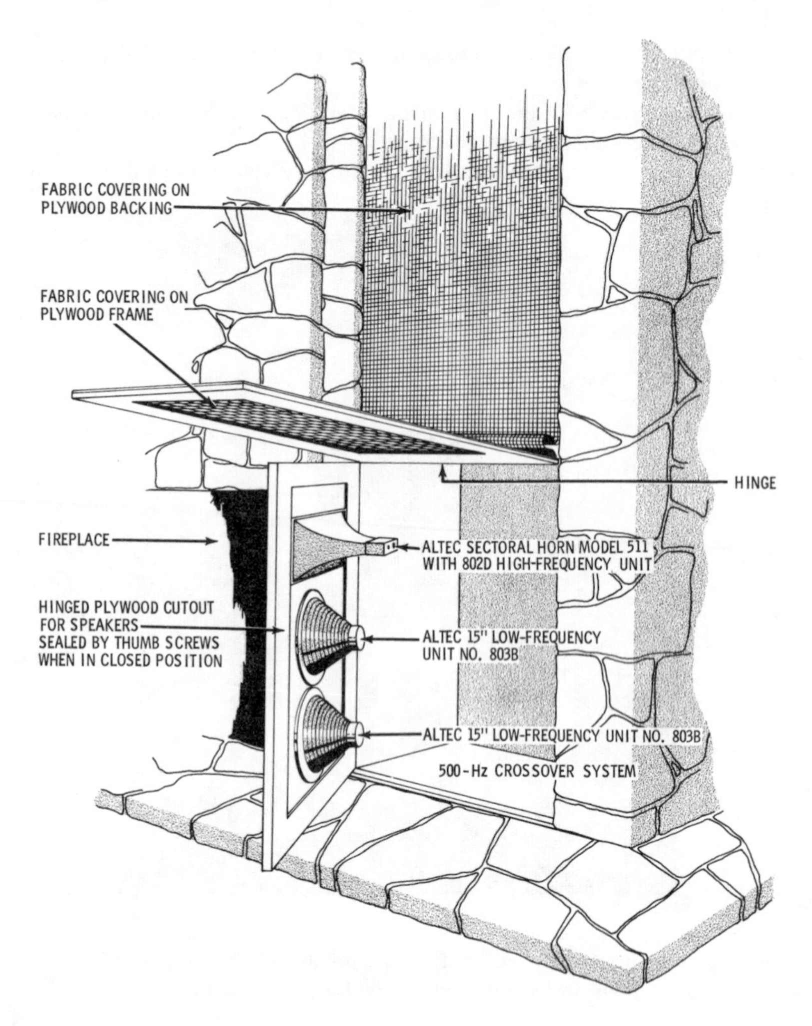

Fig. 3-3. Details of a large two-way speaker system built into a stone fireplace.

After careful consideration of all the factors, a pair of 15-inch speakers allowed a balanced compromise to be achieved in this system.

AN INFINITE BAFFLE ILLUSTRATED: SYSTEM 2

The second system illustrated is considered one of the finest designs of its type. System 2 represents the approach to an infinite baffle system that can be moved and yet meets the design requirements with an absolute minimum of compromise. System 1 requires home ownership and, for practical purposes, that the house be built with the speaker as a major objective.

System 2 is the Bozak B-310, construction plans and details of which are reprinted here through the courtesy of the R. T. Bozak Manufacturing Company of Darien, Connecticut.

Fig. 3-4 shows System 2 in a completed form. Fig. 3-5 illustrates the working plans for its construction, while Fig. 3-6 is the wiring diagram of the speakers and crossover network. Fig. 3-7 shows the arrangement of the speakers on the mounting panel.

Fig. 3-4. Infinite baffle in a movable furniture cabinet.

Courtesy R. T. Bozak Mfg. Co.

Construction of the Bozak B-310 and B-310A

The B-310A differs from the B-310 in that it has two 16-ohm Bozak B-209 midrange speakers, whereas the B-310 has a single 8-ohm B-209. Construction of the B-310 cabinet is a relatively challenging project, and will probably be undertaken only by those experienced in fine cabinetwork. Hence we shall dispense with details of construction techniques, and merely stress a few hints and cautions regarding some of the more vital operations.

Rigidity—The B-310 enclosure must be as rugged and rigid as possible. This is true of any speaker enclosure, but is especially important for the B-310 because the tremendous low-frequency power of the four Bozak B-199A bass speakers can easily cause severe cavity resonance in an inadequate structure. All joints, except the removable back panel, the speaker panel, and the back of the

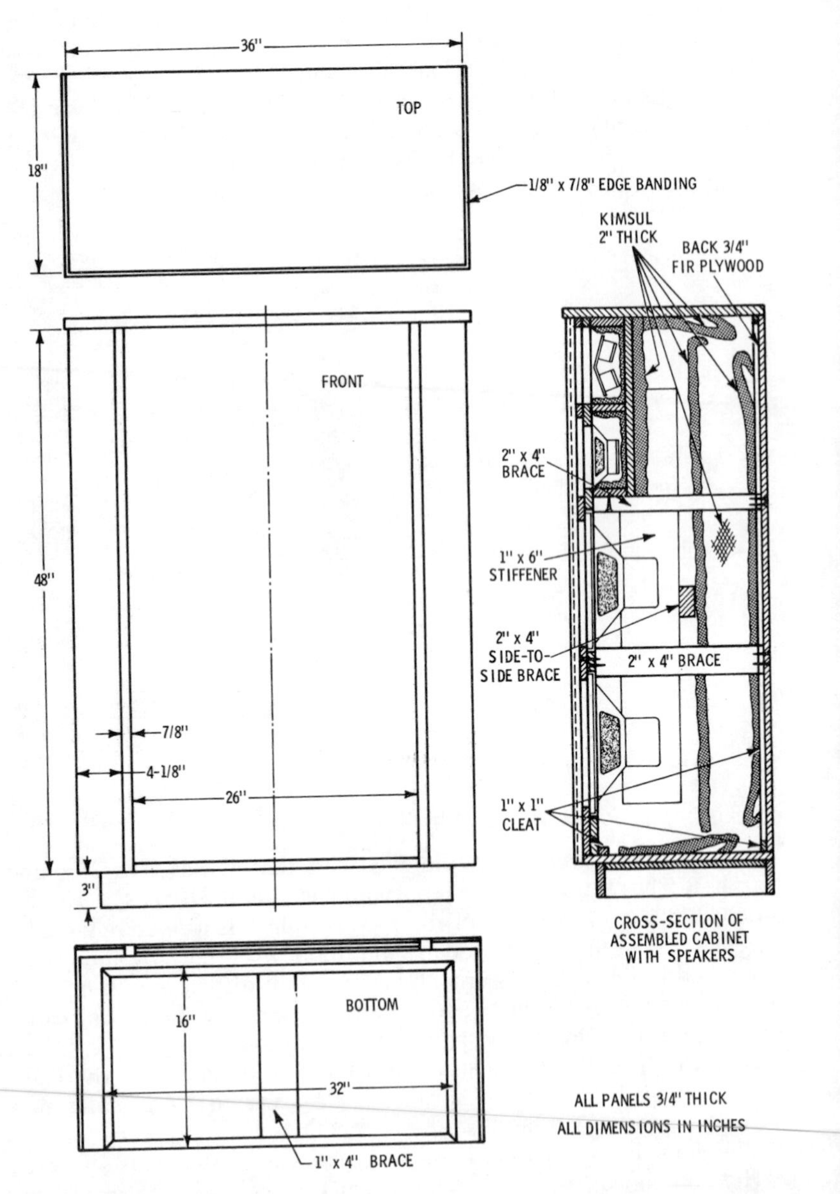

Fig. 3-5. Construction of

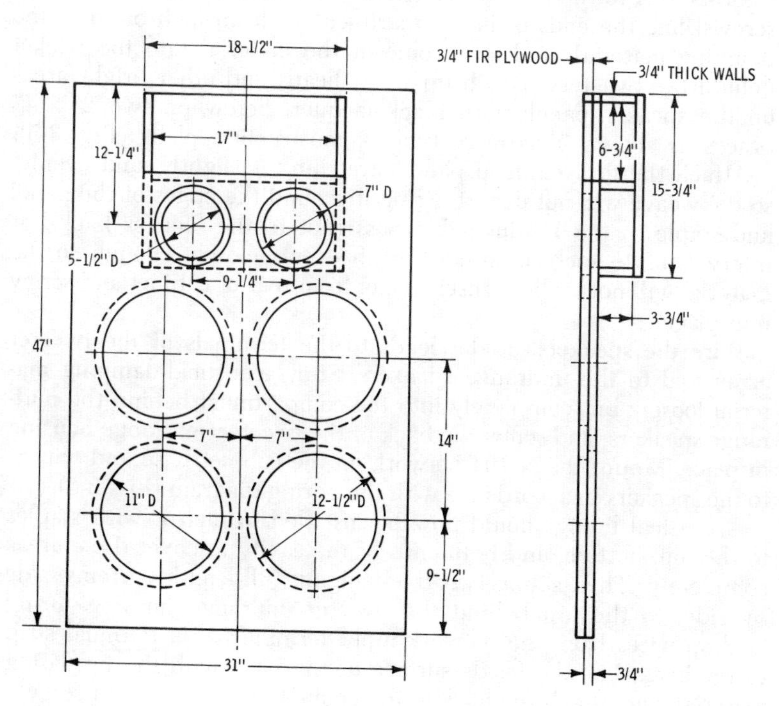

Courtesy R. T. Bozak Mfg. Co.

the B-310A speaker system.

tweeter/midrange housing, must be both glued and screwed. Where screws alone are used they should be at intervals of 6 inches or less.

Grill Cloth—This must be a nonorganic material, such as woven *Saran* or other similar plastic, to permit free transmission of the high frequencies.

Wiring—Use insulated No. 18 hookup wire, and solder all connections.

Lining—An acoustical-damping material which will not settle or pack, such as 2-inch *Kimsul* or *Fiberglas,* should be used where directed.

General Procedure—Complete the construction of the shell of the cabinet and apply the final finish before installing the speaker panel, braces, crossover network and wiring, and the acoustical lining.

Speaker Panel—Drill holes for the No. 10-32 flat-head machine screws with which the B-199A's and B-209's (one for the B-310, and cover other opening with plywood) will be attached, and apply a coat of flat black lacquer on the front of the panel. When the lacquer is dry, mount the bass and midrange speakers. Then screw the B-200XA to the back of its housing with No. 8 round-head wood screws, line the ends of its compartment with an inch or so of the damping material, and stuff some of the material into the pocket behind the tweeters. Touch up screw-heads and other bright areas on the speaker panel, with black lacquer. Screw on two 2″ × 4″ braces as shown in cross-section of construction plans (Fig. 3-5).

Attach the grill cloth to panel, stretching it slightly, and evenly, so the weave will not distort. Bring it around the edges of the panel and staple to the 1½-inch thickness (not to the face or back), at intervals of ½ inch or so, and in all events closely enough so the material will not scallop. Insert panel from rear of cabinet and screw into place.

Wire the speakers. Solder leads to the terminals of the tweeter array and to the midrange speakers. Stuff acoustical-damping material loosely but completely into the compartment behind the midrange speakers, and screw the back of the tweeter/midrange housing in place. Mount the N-104 network on the cabinet base and wire it to the speakers in accordance with the wiring diagram (Fig. 3-6).

Acoustical lining should now be installed. Attach it with staples to the top, bottom, and both ends of the cavity to cover these areas completely. Then staple two free-hanging full-length curtains, side by side, to the top behind the tweeter/midrange housing. Bring hookup wires from the network input terminals to a terminal strip to the back, line the inside surface of the back with the insulating material, and attach the back to the cabinet and braces with screws.

Your amplifier should be capable of delivering at least 50 watts of clean power at every frequency from 20 to 20,000 Hz, and all other associated equipment must likewise be of the highest quality. Damping factor: 8 to 20.

* * *

The choice in System 2 was four 12-inch drivers for the bass range. In this system, design considerations were toward excellent transient

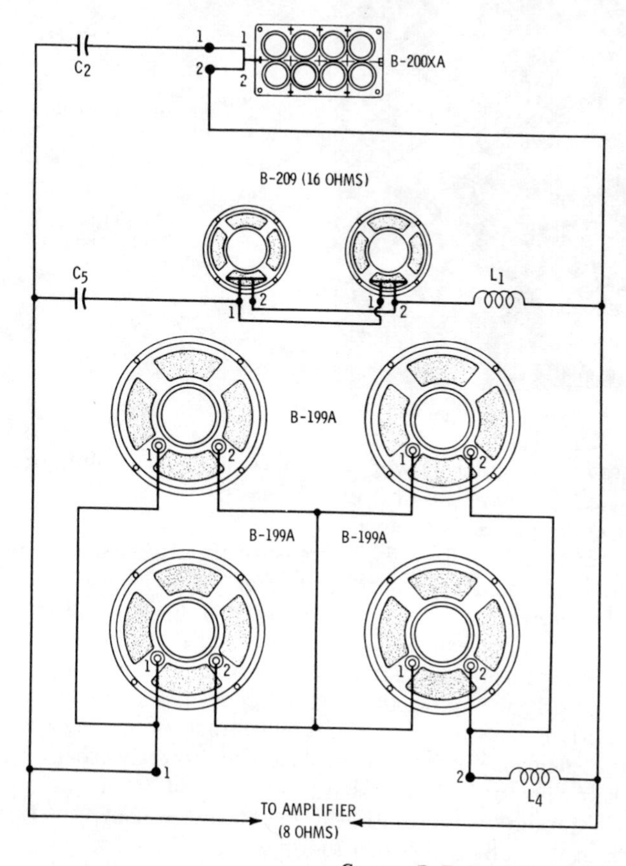

Courtesy R. T. Bozak Mfg. Co.

Fig. 3-6. Schematic for the B-310A speaker system.

response and very smooth frequency response. The efficiency of the system is somewhat below 4 percent.

Note that all cone drivers are employed. The proportion of drivers —four bass, two midrange and eight high-frequency—is classic. The high-frequency cones are splayed horizontally and vertically to achieve good distribution of high frequencies.

Fig. 3-7. An all cone-speaker system.

Courtesy R. T. Bozak Mfg. Co.

Suggested amplifier power for this system is a minimum of 50 watts at all frequencies, and experience has shown that 100 watts will not endanger the system.

The care that must be taken to ensure rigidity if stray vibrations are to be eliminated is illustrated in Fig. 3-8.

Fig. 3-9 shows the extensive bracing that is required in speaker enclosures. It is not exaggerating to say that it is hardly possible to make an enclosure too massive or over-braced. The more rigid the enclosure walls are made, the better the results achieved.

Fig. 3-10 shows the drivers and the crossover network mounted inside the enclosure before the application of the absorbent material.

The effective use of absorbent material in such enclosures to eliminate standing waves within the enclosure is shown in Fig. 3-11. This material can be sheets of *Fiberglas* or *Celludown.*

Note an extremely important detail at this point: Only the woofers are using the main enclosure—the midrange and the high-frequency units are sealed off in separate enclosures. There are several important reasons for this. First, the high-frequency radiations from the back of the high-frequency cones, if allowed to reflect off the rear of the main enclosure, would pass out through the cone of the low-frequency driver and cause phase interference. Second, the compression effect of the woofers during the reproduction of extreme bass notes could modulate the tweeter cones. It is very vital that these units each be housed separately. When a horn-type system is used as a high-frequency speaker, it is its own housing because the back of it is totally sealed.

Fig. 3-12 illustrates the detail of a well-made corner joint and suitable corner bracing.

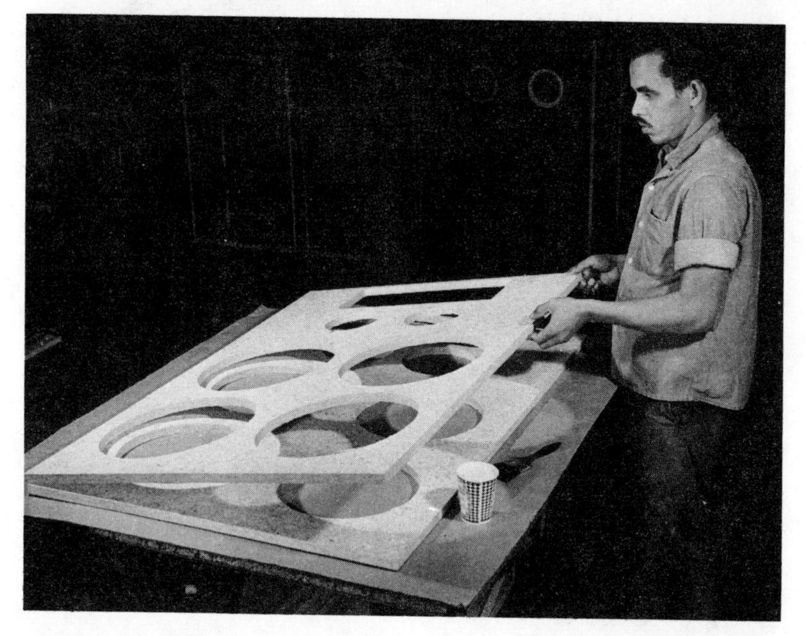

Courtesy R. T. Bozak Mfg. Co.

Fig. 3-8. Front-panel construction of the B-310A speaker system.

AN INFINITE BAFFLE ILLUSTRATED: SYSTEM 3

The final system to be considered is the greatest compromise of the three, but it illustrates the amazing results that can be achieved even when major concessions to optimum dimensions are made. As long as good engineering is employed to balance compromises, the end result can be very usable.

Cone Excursions Compared

In Chapter 2 it was mentioned that as the frequency decreased, the cone excursion increased. It was also mentioned that as the cone diameter increased, the excursion decreased. The first system discussed in this chapter employed a large-size cone for small excursions and good bass.

The second system used several smaller speakers to accomplish the same results—reducing the cone excursion by increasing the

Courtesy R. T. Bozak Mfg. Co.

Fig. 3-9. Interior view showing stiffener and cross bracing.

Courtesy R. T. Bozak Mfg. Co.

Fig. 3-10. Interior view with speakers and network in place.

Courtesy R. T. Bozak Mfg. Co.

Fig. 3-11. Interior view showing sound-damping *Celludown*.

cone area. This third system takes another tack to deal with the excursion-area conflict. Instead of increasing the cone area, the cone excursion was increased. This means that if a 12-inch speaker of the type in System 2 were to be used in System 3, the cone excursion in System 3 would have to be increased four times that of System 2. Therefore, for a 50-Hz tone the speaker in System 3 would have to

Courtesy R. T. Bozak Mfg. Co.

Fig. 3-12. Cabinet side being joined to base.

go four times the distance of the speaker in System 2 from a dead stop to both its full forward and to its full rear positions and it would have to do this in the same amount of time—1/200 of a second (1/4 of 50 Hz takes 1/200 seconds).

More energy, then, is required for System 3 because a greater mass is required to travel farther and faster. The longer distance of cone travel also gives rise to the possibility of *Doppler distortion.*

Doppler Distortion

If the cone is moving its maximum distance back and forth in the process of generating a very low-frequency tone, while simultaneously generating a high-frequency tone, the high-frequency tone is first moved forward at the rate of the low-frequency note and then backward. As the cone makes its large movement forward toward the listener, the pitch of the higher frequency riding on the cone rises. As the cone moves away from the listener, the pitch of the higher frequency lowers. This is the same effect as that heard when a train, blowing its whistle, passes by at a high rate of speed. First, the whistle is heard at a higher pitch as the train speed is added to the speed of sound, then the pitch of the whistle lowers as the speed of the train is subtracted from the speed of sound.

Fig. 3-13. Infinite-baffle enclosures, each two-cubic-feet in volume.

Advantages

If these disadvantages exist, why try to increase the cone excursion? The answer lies in the fact that by so doing the enclosure size can be drastically reduced. Enclosures as small as two cubic feet can be built with bass response down to the 35-Hz region. By using the compression of the entrapped air volume in the very small enclosure to replace part of the spring normally built mechanically into the speaker, the undesirable effect of compression can be put to work as a useful mechanical element. If a "long throw" type of driver

Fig. 3-14. Two cone drivers mounted in a two-cubic-foot infinite baffle.

is mated to a very small volume enclosure, the compression of the entrapped air will raise the natural resonant frequency of the cone.

In the course of making it a "longer throw," or longer excursion driver, the cone is made heavier in order to make it exhibit a lower open-air mechanical resonance, normally on the order of 15 to 20 Hz. The "springiness" of the enclosed air is additive with the mechanical spring which causes the resonance to rise to approximately 35 Hz.

The combination of a cone with a heavier mass, longer cone travel and air cushion brings the efficiency figure of this type of infinite baffle below 1 percent. (Chapter 2 explained that as mass increases efficiency decreases.)

Fig. 3-13 illustrates the very small size that can be achieved, and Figs. 3-14, 3-15, and 3-16 show the interior construction. Again, it must be constructed of heavy materials, adequately braced, and securely fastened. The lavish use of acoustical absorption material performs the same function as increasing the volume of the enclosure by a factor of 1.4. The absorption material helps convert the pressure-temperature changes from adiabatic to isothermal.

SUMMARY

Infinite baffles, it can be seen, dictate the choice of driver for a given size of enclosure. Infinite baffles allow experimentation with the basic parameters of speakers. Like all other engineering projects,

Fig. 3-15. Absorbent material being installed in the enclosure.

Fig. 3-16. Enclosure completely filled with absorbent material.

extremes must be brought into a balanced compromise for the most usable results. Within the limits of the internal volume chosen for the enclosure, it may be built in any shape decor dictates except that any two dimensions should not exceed a ratio of one to three. Choice of volume is, of course, decided by the largest physical size which you can tolerate.

The basic problem encountered in using the infinite-baffle enclosure is "grabbing hold of enough air" at very low frequencies. A speaker can be likened to a canoe paddle. As the frequency decreases, the paddle slowly changes to a small teaspoon. The remedy is to either use a bigger paddle in the first place, or more paddles, or to sweep farther with each stroke of the paddle. Wouldn't it be nice if someone could make a paddle that grew larger instead of smaller as the frequency decreased? Well, someone could, and someone did, which leads to the subject of the next chapter—bass-reflex enclosures.

4

Bass-Reflex or Phase-Inversion Enclosures

Bass-reflex enclosures (or *phase-inversion* enclosures, as they are more correctly referred to) offer the best approach to utilizing the energy generated by the back radiation from the cone speaker. If the cone is not baffled, the back radiation interferes with the front radiation. Infinite baffles ensure that the back radiation never meets the front radiation in a detrimental manner.

BASS REFLEX DEFINED

The phase-inversion enclosure allows the back radiation to be usefully added to the output of the front radiation. This improves the low-frequency response of the system. The mechanics of this process may be described as follows: The entrapped air volume of the enclosure is used as an extension of the cone to move a volume of entrapped air in a port or opening approximately equal to the area of the air displaced by the front of the cone.

The air volume in the enclosure acts as a "spring" coupling the back of the cone to the "cone" of air entrapped in the port area. This "spring" has the effect of delaying the transfer of the cone movement to the volume of entrapped air in the port area. This

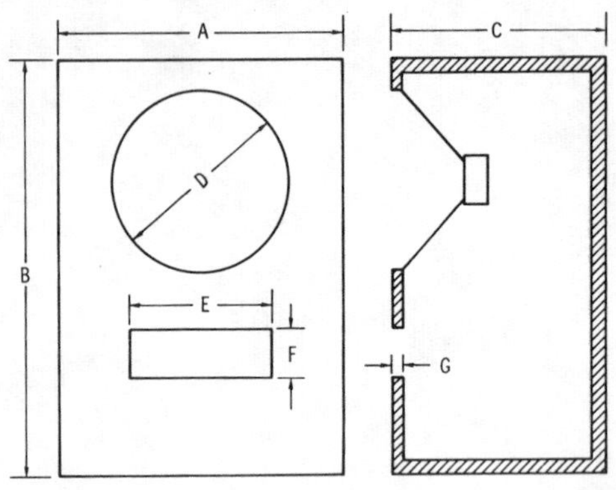

Fig. 4-1. Basic dimensions of phase-inversion enclosures.

time delay brings the two radiations (from the cone and from the port) from an out-of-phase, or opposed, condition into an in-phase condition at the necessary low frequency.

One popular misconception regarding phase inverters is that as the frequency decreases, the front radiation from the cone becomes less, and all the acoustic power heard emerges from the port. Actually, the very best that can ever be achieved would be approximately equal outputs from the port area and from the front of the cone, inasmuch as radiations from the port are directly related to the movement of the cone. One cannot get "something for nothing." As a matter of fact, there is slightly less acoustical power from the port area as compared to the front of the cone because some energy is lost via absorption inside the enclosure.

Fig. 4-1 illustrates the main features of a phase-inversion enclosure. Dimensions $A \times B \times C$ (in inches) = the volume of the enclosure in cubic inches. For conversion to cubic feet divide cubic inches by 1728 ($=12^3$). D is the speaker mounting opening, which is equal to the diameter of the surround compliance of the driver. $E \times F$ is the port area; $E \times F \times G$ is the port volume.

With this type of enclosure, three main variables are adjusted to achieve increased low-frequency response: (1) enclosure volume can be altered, (2) port area and volume can be scaled up and down, and (3) the free-air resonance of the speaker cone may be of different values for various sizes of cones and types of drivers.

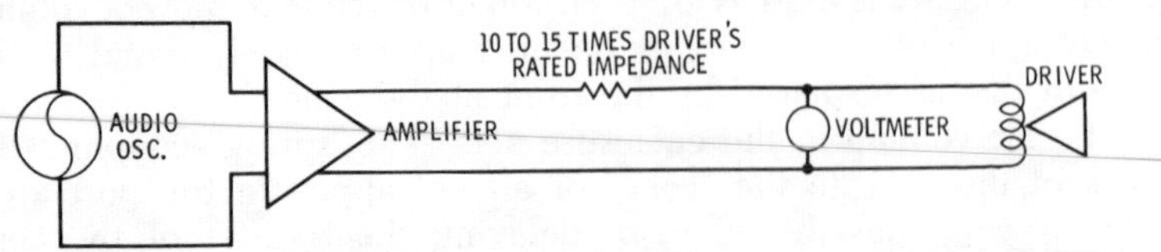

Fig. 4-2. Electrical circuit for measuring impedance (or resonance) of a speaker.

Table 2-1 in Chapter 2 states the rated diameter and effective piston area of the most commonly used sizes of woofers (low-frequency speakers). It is important to know the effective piston area of the cone used because, ideally, the port should be the same area. If the port is made larger than the cone area, there is danger that the speaker is no longer baffled. At this point the port becomes large enough to allow interfering back radiations to mingle with

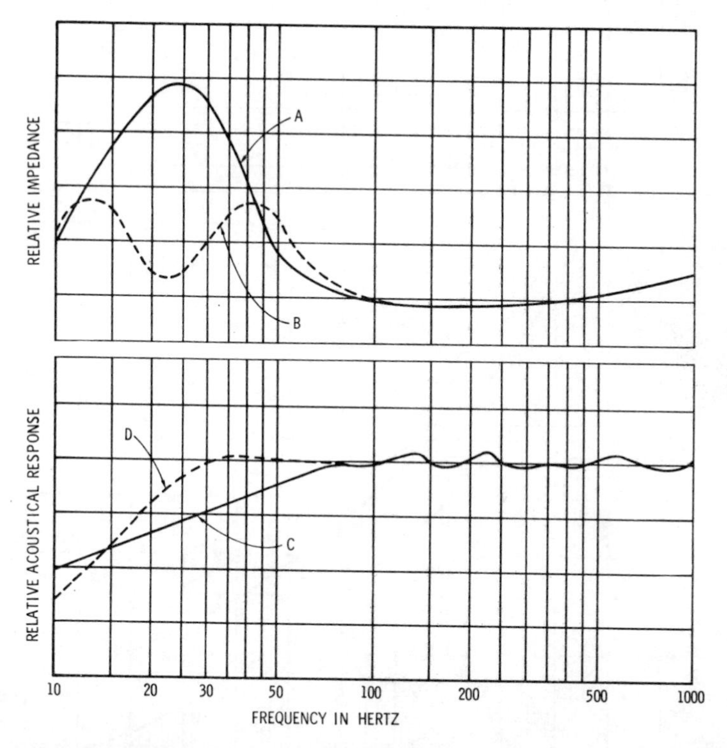

Fig. 4-3. Characteristic curves for a low-frequency driver. (A—impedance of driver in free air, B—impedance of driver in a tuned enclosure, C—acoustical response of driver mounted in an infinite baffle, D—acoustical response of driver mounted in a ported enclosure).

front radiations as the frequency increases above resonance. If the port is made smaller than the cone area, then the port radiations cannot equal the output of the front radiations from the cone. When the cone area and the port area are equal, they can generate approximately equal acoustical energy, which results in almost double the output that would be expected from the same driver in an infinite baffle at the same low frequency.

MEASURING THE CONE RESONANCE

The speaker chosen must be measured to find its free-air resonance. The only tools needed are an inexpensive audio oscillator, a voltmeter, and a resistor having a resistance value between ten and 50 times the value of the rated impedance of the speaker (see Fig. 4-2).

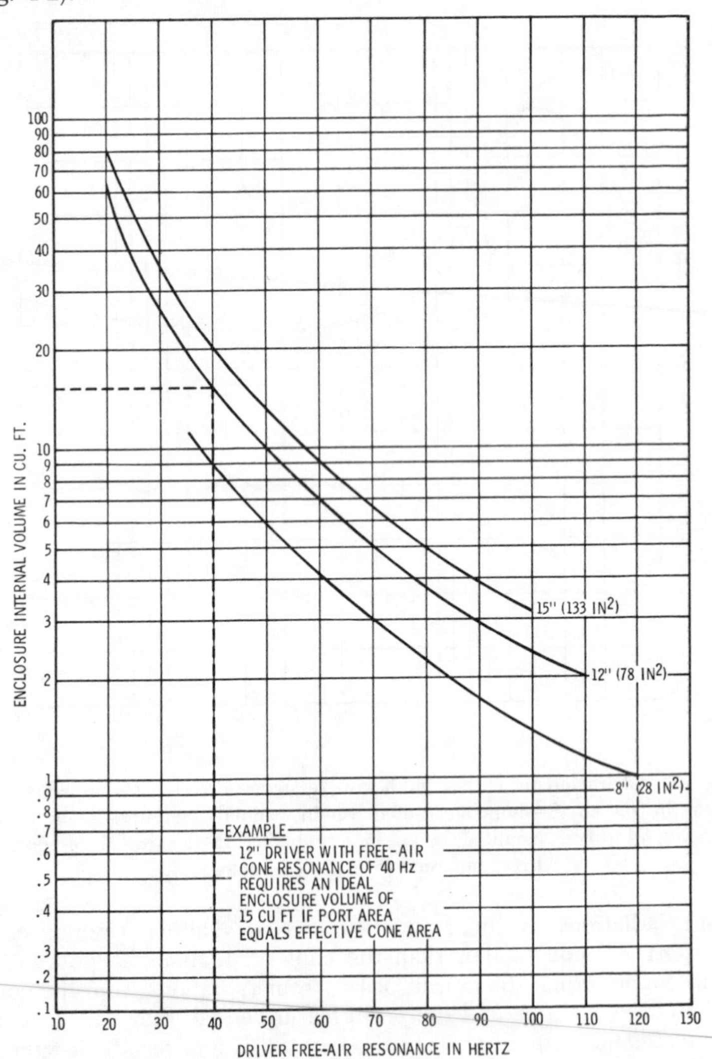

Fig. 4-4. Ideal relations between enclosure volume and driver resonance when the port area is equal to the effective cone area.

The speaker, without an enclosure, should be held up in the air, free of obstructions and possible reflections. As the oscillator is gradually adjusted to a lower frequency, the voltmeter reading will increase as resonance is approached. The peak reading obtainable on the voltmeter is the point of resonance. The frequency on the dial of the oscillator at this reading is the resonant frequency.

The lower the free-air resonance of the speaker, the larger the enclosure will have to be to exhibit the same resonance if a port area equal to the cone area is also achieved. Actually, as the free-air resonance is approached, the electrical phenomenon that causes the voltage rise in the voltmeter is the increasing impedance (complex) of the speaker as the frequency is lowered (see Fig. 4-3).

If the speaker is rated at 16 ohms and matched to an amplifier of the same nominal rating, then, "as the impedance grows, the wattage goes." This is Ohm's law operating. As the "load" impedance increases, less power is drawn from the amplifier due to mismatch.

Knowing the free-air cone resonance of the speaker (see Fig. 4-4) enables one to find the volume of the enclosure that will resonate

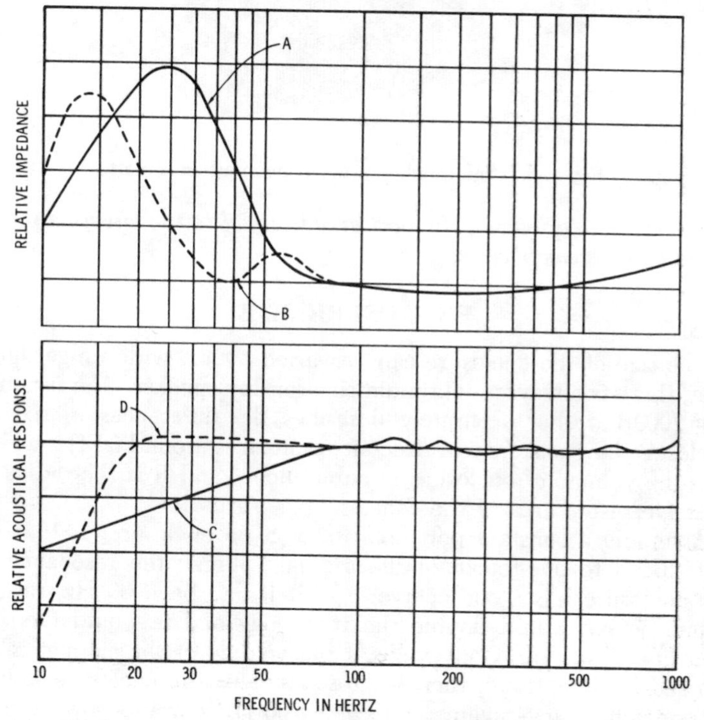

Fig. 4-5. Varying the response by changing the enclosure resonance.

Fig. 4-6. High-frequency horn mounted in a port.

at this same frequency *if the port area is equal to the effective piston area of the speaker cone.*

ENCLOSURE SIZE

The size of the enclosure can be varied over a wide range; however, there is always an optimum size for any speaker. The formula, $V = 2000R$, rather accurately describes the differences in size of enclosure for speakers with free-air resonances around 60 Hz, where V is the volume of enclosure in cubic inches and R is the radius of effective piston area of the cone.

Assuming a constant port area for a given enclosure, as the size is reduced to one-fourth of its original volume, the resonance of the enclosure raises one octave; e.g., if it had been 60 Hz, raising it one octave would double the frequency and it would then be resonant at 120 Hz. Conversely, if the volume of the cabinet is increased by four times, then the resonance of the enclosure lowers by one octave; e.g., again, if originally 60 Hz it would now be one-half, or 30 Hz.

If it becomes necessary to sacrifice some bass efficiency (thereby reducing the size of the cabinet) to better adjust to demands of decor, the formula to use is:

$$R = \frac{V_2}{V_1}$$

where,

R = ratio of port area change,
V_1 = volume of the enclosure with port area equal to cone area,
V_2 = desired volume of the smaller enclosure.

And if,

P = port area of optimum size enclosure,

then,

R^2P = new port area for smaller enclosure.

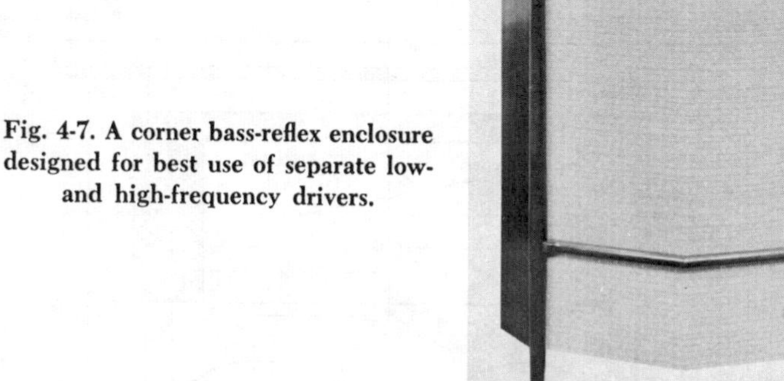

Fig. 4-7. A corner bass-reflex enclosure designed for best use of separate low- and high-frequency drivers.

If the enclosure were 10 cubic feet with a 12-inch speaker (free-air cone resonance 50 Hz) and a port area of 78 square inches, to reduce the enclosure to five cubic feet, calculate:

$$R = \frac{5 \text{ cu ft}}{10 \text{ cu ft}} = 0.5$$

$$R^2P = (0.5)^2 \times 78$$
$$= 0.25 \times 78$$
$$= 19.5 \text{ sq in}$$

While this reduced the size of the enclosure and the "tuning" remained at the same frequency, it was at the expense of bass output. There is a loss of bass output due to the greatly reduced port area

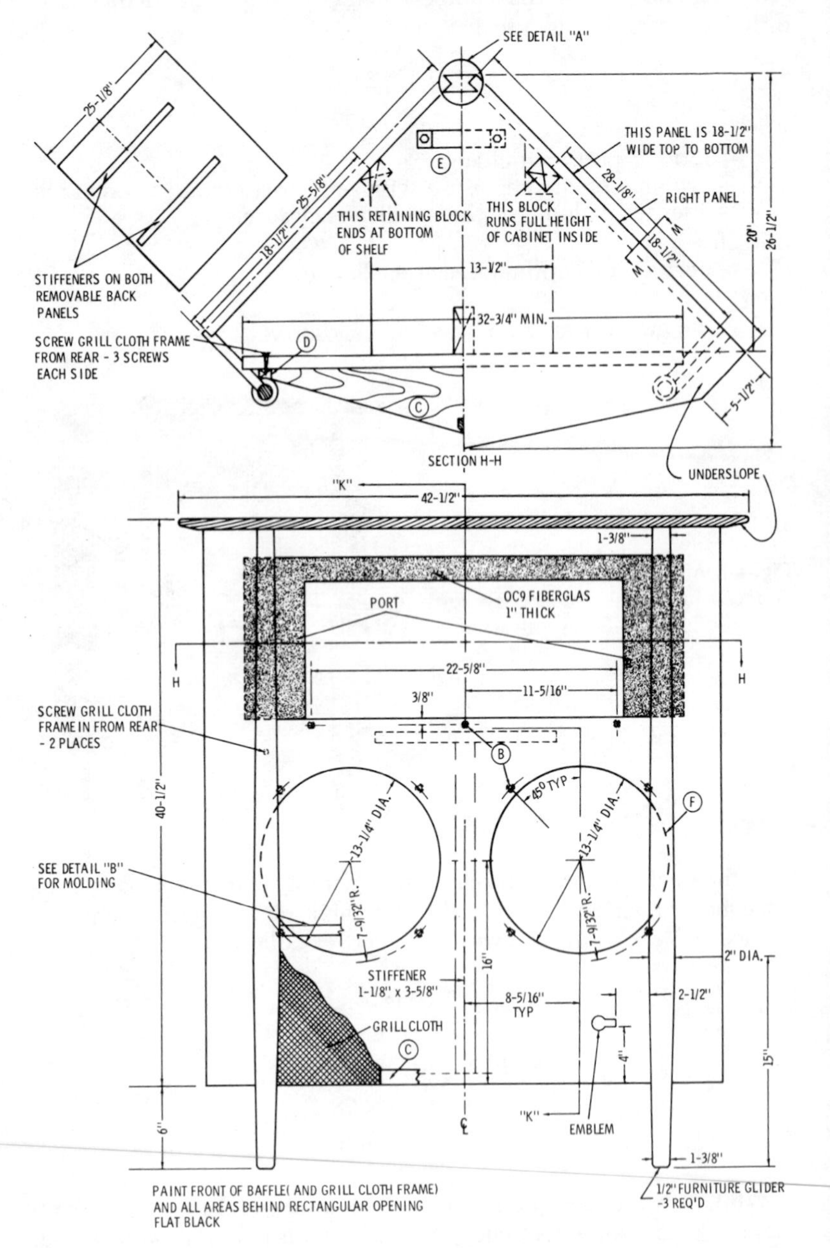

Fig. 4-8. Construction drawings for

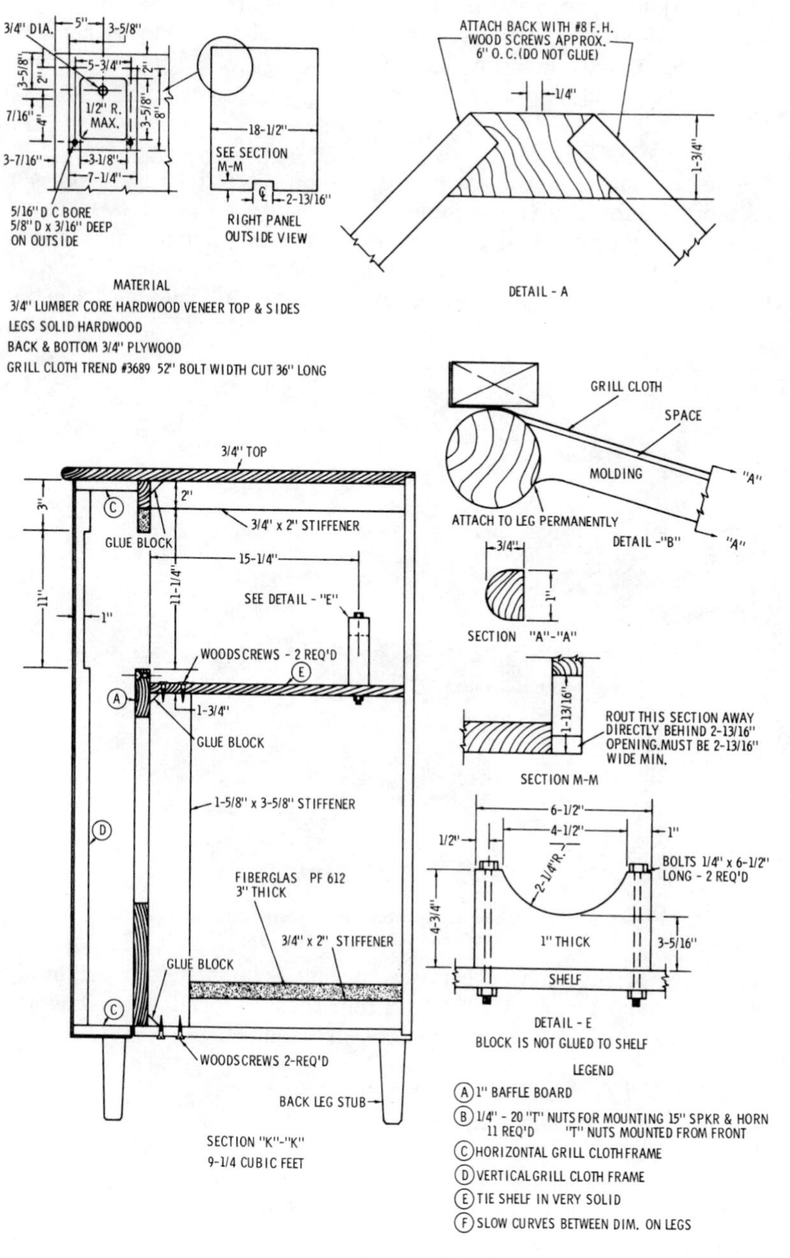

Courtesy Altec Lansing, Div. LTV Ling Altec, Inc.

the enclosure shown in Fig. 4-7.

compared to cone area, as well as increased low-frequency distortion caused by the increased movement of the cone as a result of the decreased enclosure loading.

PORT SIZE

The resonance of the enclosure can be altered by still another method. If, for a given volume of enclosure, the port is reduced to one-fourth its original area, the resonance of the enclosure lowers by one octave; e.g., if originally 60 Hz, it now becomes 30 Hz. If the area of the port is increased to four times its original area, the

Fig. 4-9. Bass-reflex enclosures in a stereophonic system.

resonance of the enclosure raises by one octave; e.g., if originally 60 Hz, it now becomes 120 Hz. From these interrelations it is apparent that for a given free-air cone resonance:

1. A small enclosure requires a small port area.
2. A large enclosure requires a large port area.

As mentioned earlier, the ideal port area should equal the cone area of the speaker. The preceding relation shows how a smaller than optimum port area can be utilized in order to allow reduction in the size of the enclosure. On the other hand, even if size were of no consequence, designing the port area larger than the speaker-

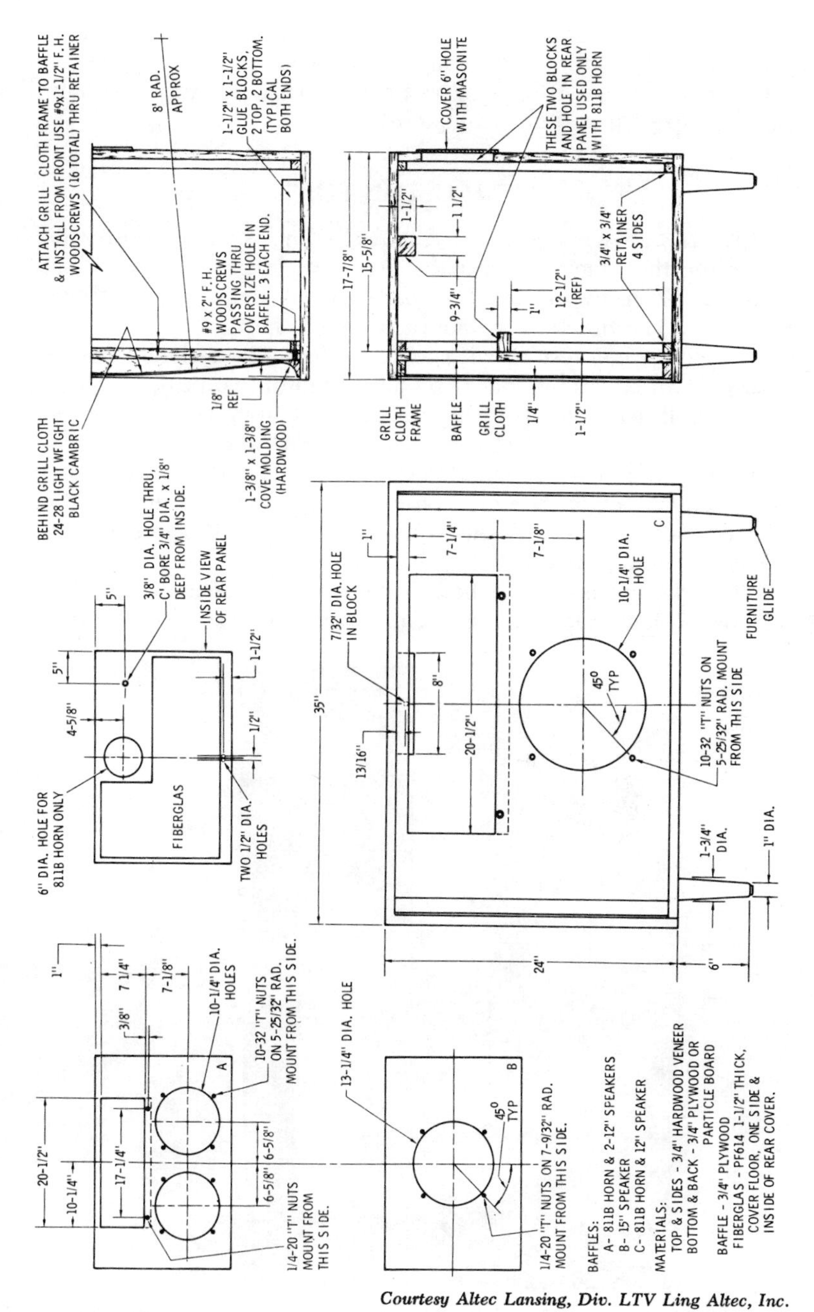

Courtesy Altec Lansing, Div. LTV Ling Altec, Inc.

Fig. 4-10. Construction drawings for the enclosure shown in Fig. 4-9.

cone area only results in unbaffling the woofer. The effect of a port area larger than the cone area is most noticeable at frequencies other than the low-frequency resonance. This is often overlooked even by speaker system manufacturers.

TUNING THE PORT

After design work is finished and the enclosure is constructed, it is often the practice to cut the port area larger than the design calls for. Then, by means of a board moved across the port opening, a final careful "tuning" can be accomplished.

Using the same test circuit shown in Fig. 4-2, gradually "sweep" the oscillator across the lower frequencies until two peaks are found on the voltmeter. Keep changing the port area with the sliding board, while sweeping the low frequencies, until the two peaks previously found are equal in amplitude. The enclosure is now tuned, and proof of its correct adjustment is that the dip between the two peaks is now at the same frequency as the free-air cone resonance of the driver.

When the enclosure is tuned to the same frequency as the free-air cone resonance of the speaker, the impedance of the speaker

Fig. 4-11. A large enclosure made into a piece of furniture.

Fig. 4-12. A large enclosure which requires only three square feet of floor space.

mounted in the enclosure looks like Curve B in Fig. 4-3. Note that one high resonance peak has been traded for two smaller ones. The higher-frequency peak is essentially of useful phase, and the lower frequency one is of detrimental phase. However, the lower one is at a frequency so low as to be of no consequence.

Curve D in Fig. 4-3 shows the frequency response (idealized) of a driver and ported enclosure associated with the impedance Curve B. Curve C is for the same driver mounted in an infinite baffle.

One further improvement results from deliberately raising the resonant frequency of the enclosure to a point higher than the free-air cone resonance of the speaker. If the driver has a free-air cone resonance of 30 Hz, the enclosure resonance is raised to 45 Hz. The resonance of the speaker in the enclosure now looks like Curve B in Fig. 4-5. The low-frequency peak (18 Hz) is much larger, but at this low frequency, the peak can have no effect on the performance of the system. Now the high-frequency peak is quite small and the bass response is smooth and efficient all the way to the cutoff

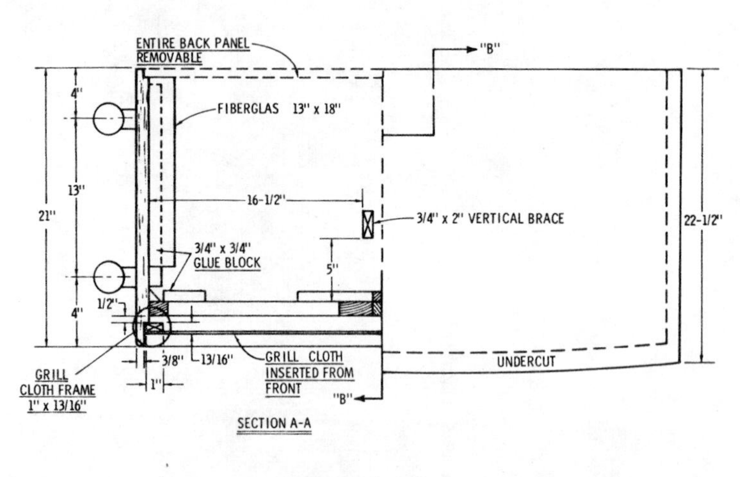

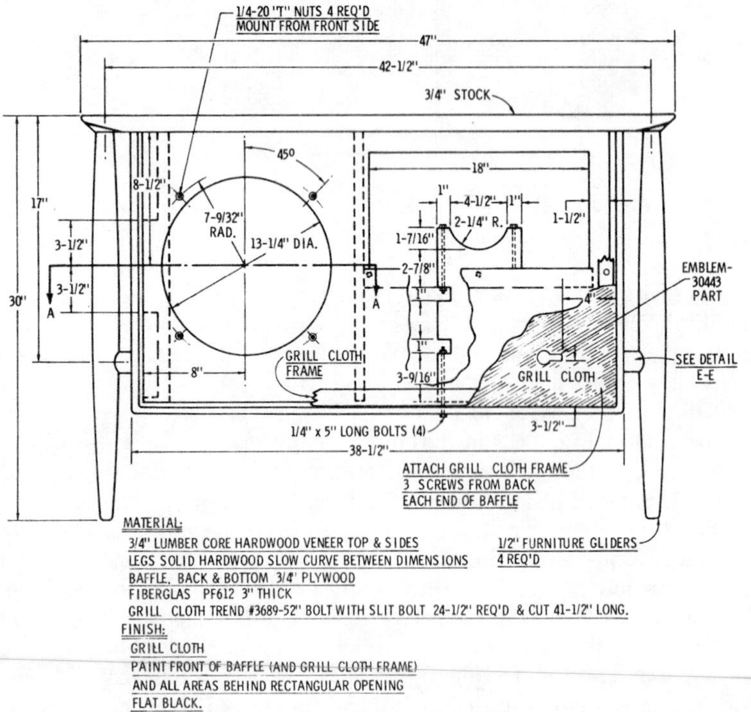

Fig. 4-13. Construction drawings for

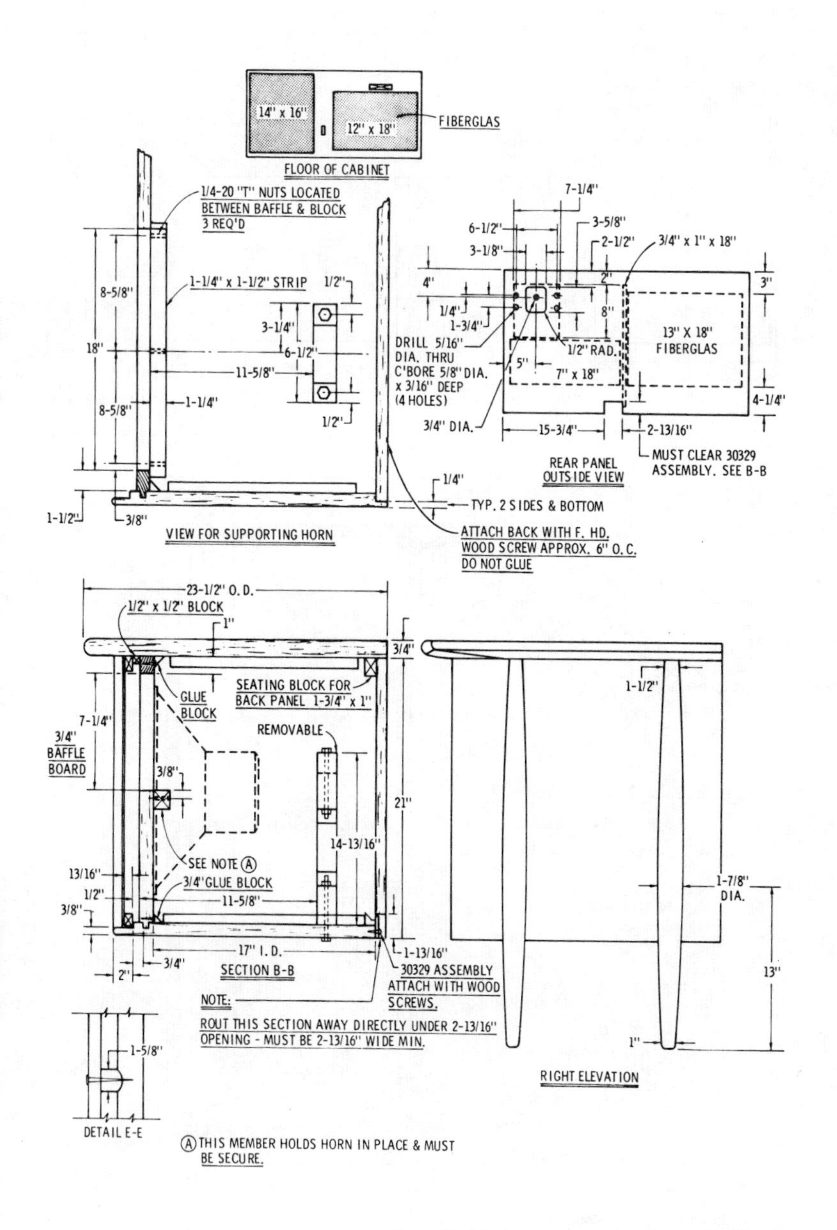

Courtesy Altec Lansing, Div. LTV Ling Altec, Inc.

the enclosure shown in Fig. 4-11.

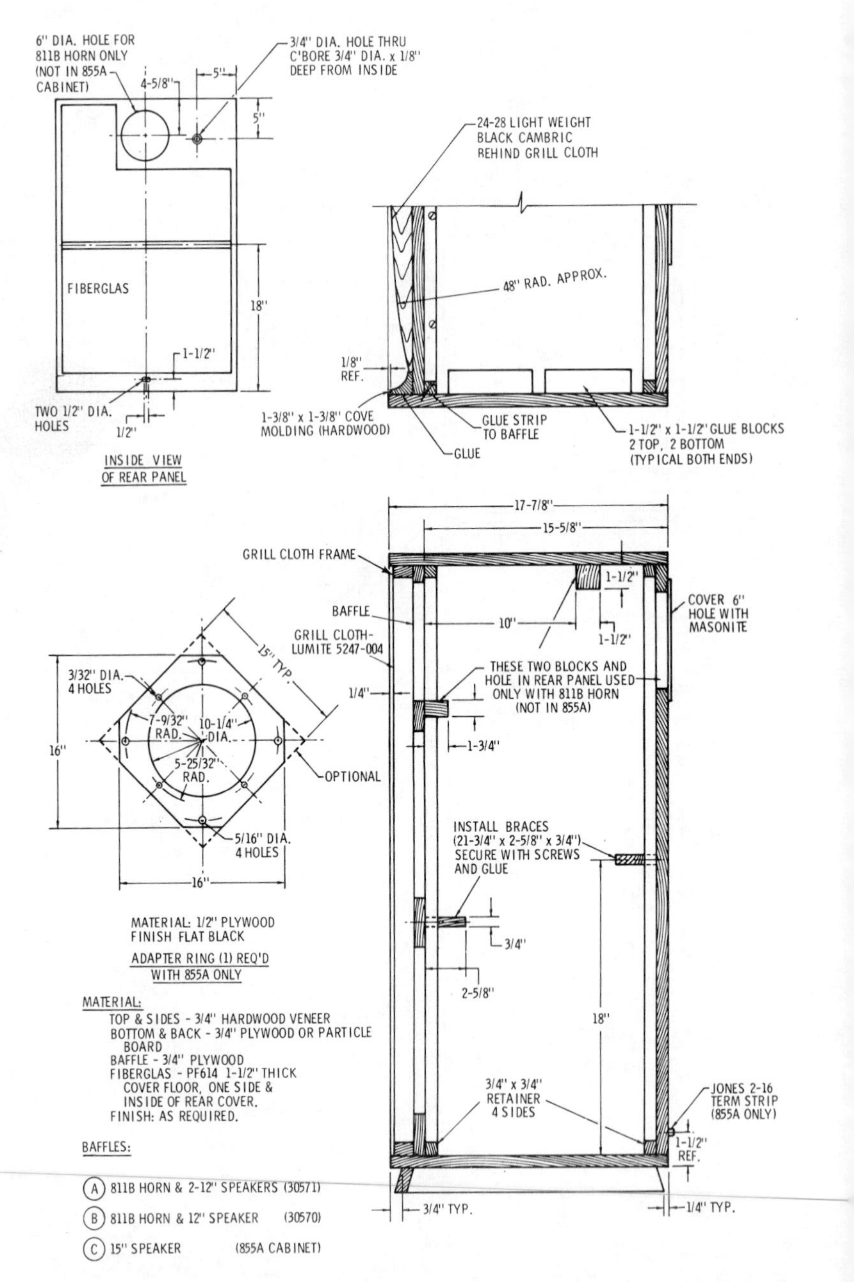

Fig. 4-14. Construction drawings for

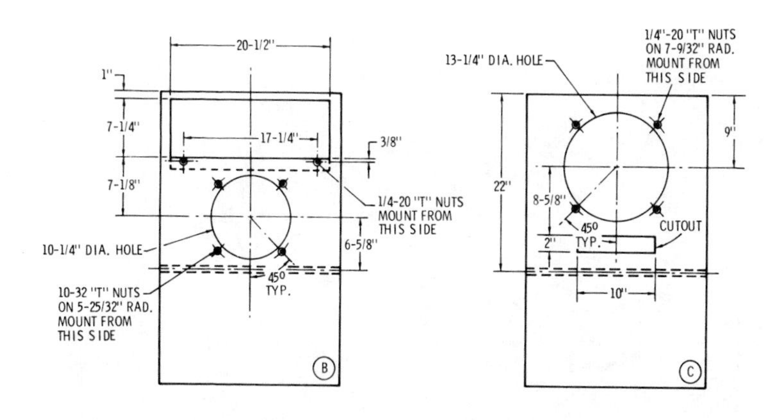

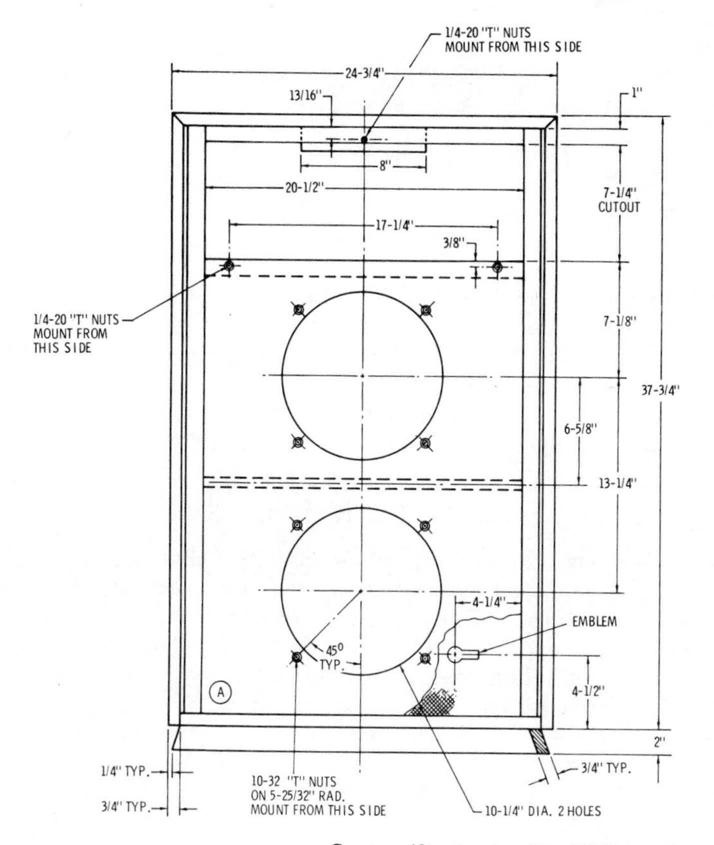

Courtesy Altec Lansing, Div. LTV Ling Altec, Inc.

the enclosure shown in Fig. 4-12.

Fig. 4-15. Bass-reflex enclosures made smaller than optimum to meet room decor requirements.

point. (See Curve D, Fig. 4-5.) This means that the speaker system absorbs more energy from the amplifier with the port tuned to a higher frequency; and as a result, a better, smoother, and more solid bass response is obtained.

Still another bonus is improved transient response due to better electrical damping in the all-important region of cutoff to 100 Hz.

CONSTRUCTION PLANS

In a number of the construction drawings that appear in this chapter, the port has been cut much larger than normally required. This is to permit the high-frequency horn and driver to be mounted within the opening. The difference between the large port area and the space the horn occupies is the actual port size. Therefore, it is this differential area that is considered as the actual port. The result is a port of the correct area surrounding the high-frequency horn (see Fig. 4-6).

The accompanying photographs and construction drawings are of thoroughly tested and proven bass-reflex enclosures. Careful adherence to the dimensions and details given will result in highly satisfactory performance (see Figs. 4-7 through 4-19).

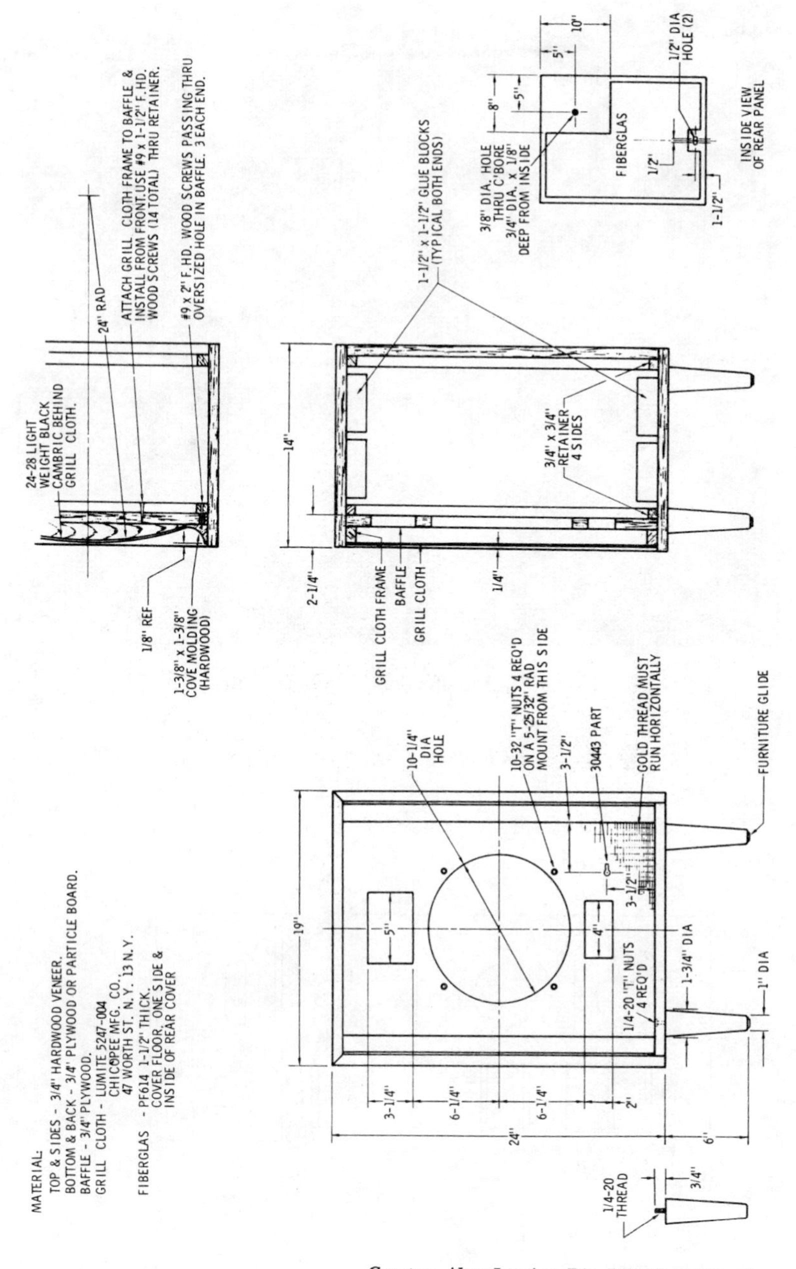

Courtesy Altec Lansing, Div. LTV Ling Altec, Inc.

Fig. 4-16. Construction drawings for the enclosure shown in Fig. 4-15.

Fig. 4-17. A professional studio monitor.

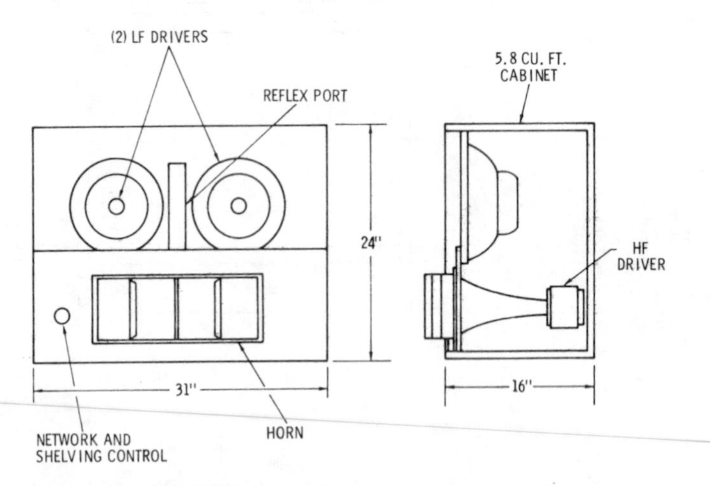

Fig. 4-18. Basic dimensions for the enclosure shown in Fig. 4-17.

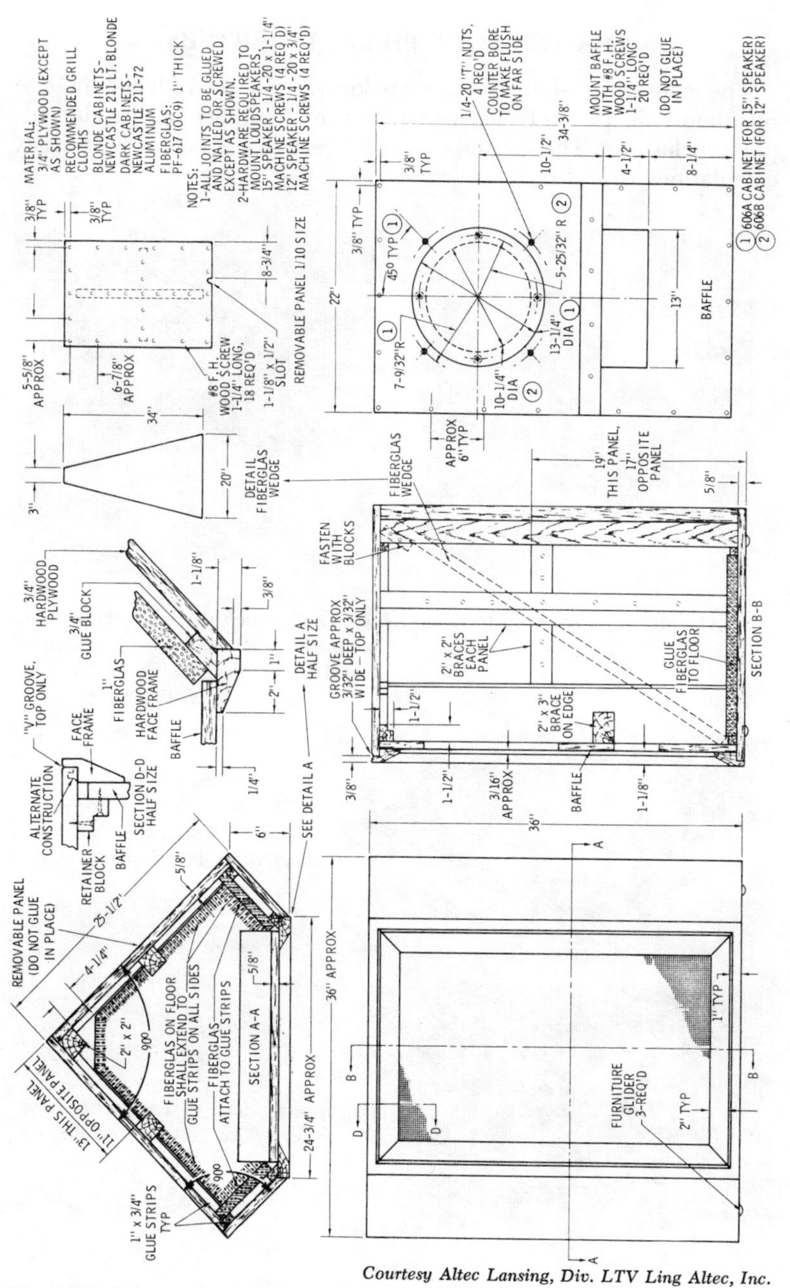

Courtesy Altec Lansing, Div. LTV Ling Altec, Inc.

Fig. 4-19. Construction drawings of a no-compromise, corner bass-reflex enclosure designed for 15-inch duplex drivers.

DUCTED-PORT PHASE INVERTERS

The conventional bass-reflex enclosure can be reduced in size by almost 50 percent, according to the designers of the ducted-port enclosures. The formula (theoretical) that governs bass-reflex calculations is:

(A) *Finished speaker system.*

(B) *The 12-inch woofer is shown here mounted on ¾-inch composition board front panel along with cone tweeter and the duct.*

(C) *A thickness of* Fiberglas *is stapled to board for damping.*

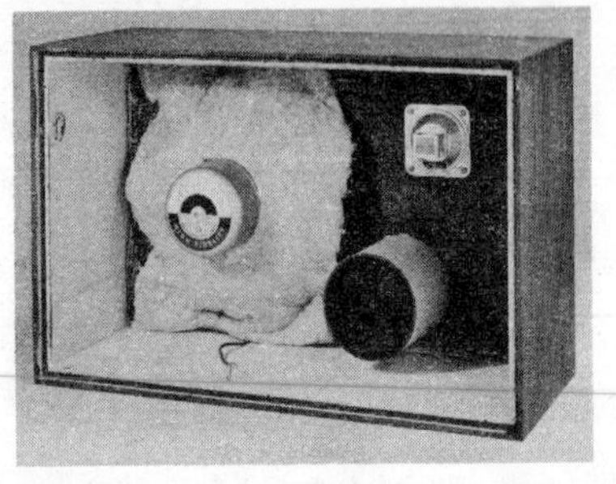

(D) *Inside view of enclosure shown before back cover is put on.*

Fig. 4-20. A high-fidelity speaker system of optimum volume that has been constructed using the data obtained from nomograms.

$$F_c = K \frac{1}{\sqrt{LC}}$$

where,

F_c = the low-frequency resonant point,
K = all the lumped constants,
L = mass of air in the port,
C = the compliance (springiness) of the entrapped air in the enclosure.

Without going into a great deal of detail (because the complex nature of K does not allow a simple value to be given to it), it can be seen that an increase in the magnitude of L allows a decrease in the magnitude of C.

This means that through the use of a duct (illustrated in Fig. 4-20) the port volume can be substantially increased.

In addition to the increased port volume it also provides better baffling isolation to higher frequencies which do not like to turn corners. Because of the very complex nature of K in the equation, plotted charts are used to calculate dimensions for standard bass-reflex enclosures.

DESIGNING A DUCTED-PORT ENCLOSURE

James F. Novak, senior engineer, Jensen Mfg. Div., The Muter Company, has researched this subject thoroughly. The following material is reprinted with his permission.*

Optimum Enclosure Size

"A careful study of bass-reflex operation from a theoretical standpoint reveals a complexity far greater than the extremely simple construction would indicate. A number of generally unknown facts come to light concerning low-frequency extension and optimum transient response.

"1. *A bass-reflex enclosure can be too large.* After a certain maximum volume is reached, further volume increases result in "boomy" bass rather than an appreciable extension of low-frequency response. This is particularly true when a speaker with a small magnet is used. It is best to use a completely closed cabinet in this case.

"2. *A bass-reflex enclosure can also be too small.* A common assumption is that the enclosure can be made much smaller by using a duct behind the port. While the use of a duct allows one to tune the enclosure to very low frequencies when volumes are small, it is the ratio of enclosure air stiffness to speaker stiffness that deter-

---

* Reprinted in part from January, 1966, issue of *Electronics World.*

mines the low-frequency cut-off. As the enclosure volume is made smaller, the enclosure stiffness increases and so does the cut-off frequency. This will remain true regardless of how the enclosure is tuned. A false "boomy" bass will generally result and the enclosure may as well be left closed. In some cases a better sounding bass will result if the back of the speaker enclosure is removed.

"3. Only one condition of cabinet tuning and damping will result in optimum transient response, *i.e.*, freedom from hangover or boom. Because a speaker in a bass-reflex forms a closely coupled system of two tuned circuits, the system will tend to decay with two frequencies. The achievement of optimum transient response demands that the system decay with one frequency and with a specified time constant. In order to make the system decay in one frequency, the enclosure volume and port combination must be tuned to the free-air resonance of the speaker. A specified decay can be achieved by adding acoustic resistance, in the form of damping material, to the enclosure. The proper amount of damping material can be determined by means of a test method described later.

"A nagging question in the design stage of any enclosure of this type is "How large shall it be?" It was pointed out earlier that the enclosure can be too large or too small for proper bass-reflex action. This implies that an *optimum volume* exists and indeed it does. This optimum volume does not depend upon the size of the speaker nor its resonant frequency *per se* but rather on the ratio of enclosure air stiffness to the speaker cone suspension stiffness. This optimum ratio is 1.44 or, looking at it another way, the speaker resonant frequency in the enclosure before porting should be 1.56 times the free-air resonance of the speaker. This size enclosure, when properly tuned, yields at the same time the most extended low-frequency response and a transient response with subjectively unnoticeable hangover, assuming sufficient damping exists. Compared to the entirely closed cabinet, the half-power point (3 db down) occurs at 0.7 times the closed cabinet speaker resonance for an extension of one-half octave.

Designing the Enclosure

"In order to proceed with the actual design work, it is necessary to know the stiffness of the cone suspension. Since speaker manufacturers are notorious for not having this information readily available, it is necessary to derive this by measuring the speaker resonance in free air and in a "standard volume." A properly calibrated audio oscillator, a simple a.c. vacuum-tube voltmeter, and a 100- to 1000-ohm resistor are required. Although this value is not at all critical, the higher values will give more sharply defined readings. Use the largest value consistent with the oscillator output volt-

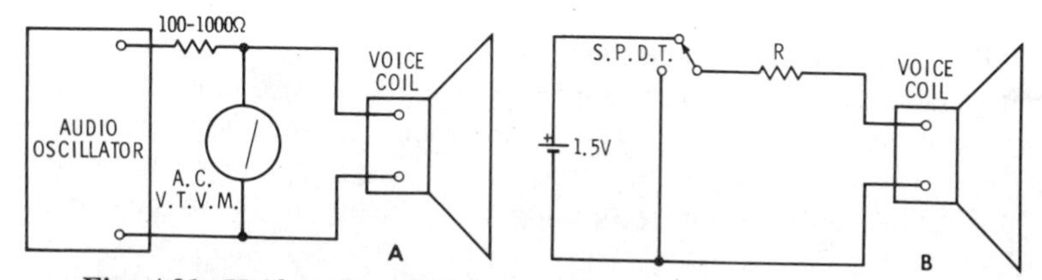

Fig. 4-21. Hookup for checking speaker resonance and damping.

age and voltmeter sensitivity. Fig. 4-21A shows how these elements are connected.

"The "standard volume" is nothing more than a small plywood box of known volume. Fig. 4-22 shows the constructional details of boxes for 8-inch, 12-inch, and 15-inch speakers. Although a single volume could have been used for all loudspeaker sizes, three separate volumes were chosen in the interest of economy. This box must be thoroughly sealed with caulking compound or putty to prevent leaks. Note: All measurements will be made with the speaker mounted on the *outside* of the box. All six sides should, therefore, be permanently assembled.

"The first step after selecting a loudspeaker and constructing the appropriate "standard volume" is to measure resonant frequencies. Hook up the unbaffled speaker as shown in Fig. 4-21A. The speaker should be held in the air away from any large objects and the audio

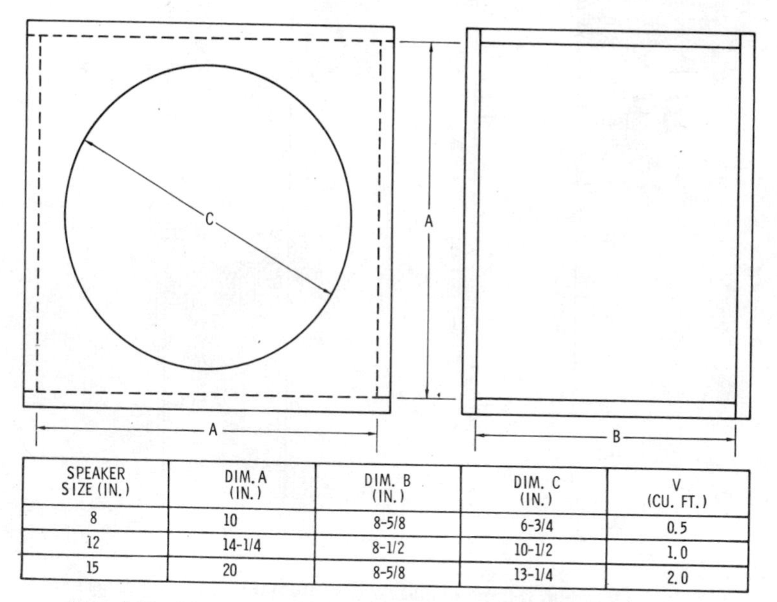

SPEAKER SIZE (IN.)	DIM. A (IN.)	DIM. B (IN.)	DIM. C (IN.)	V (CU. FT.)
8	10	8-5/8	6-3/4	0.5
12	14-1/4	8-1/2	10-1/2	1.0
15	20	8-5/8	13-1/4	2.0

Fig. 4-22. Dimensions of "standard-volume" loudspeaker boxes.

oscillator slowly swept through the low-frequency end of the audio range so that it passes through the speaker's resonant frequency. The voltmeter connected across the voice coil terminals will show a large rise in voltage at this frequency. The frequency corresponding to maximum voltmeter reading is the resonant frequency of the speaker.

"After noting the free-air resonant frequency, place the speaker face down over the hole in the "standard box." A slight amount of hand pressure should be applied to the rear of the speaker to help get a good seal between speaker gasket and the box. The speaker resonant frequency is again determined as before. The resonant frequency determined this time will be higher than the free-air resonance. It is quite possible for this frequency to be two to four times the free-air resonance.

"The proper nomograms of Figs. 4-23, 4-24, or 4-25 can now be used to determine the proper enclosure volume. The following example will clarify the technique.

"Assume an enclosure is to be built for a 12-inch loudspeaker with the following resonant frequencies: free-air resonant frequency, 62; resonant frequency in "standard box," 121 cps.

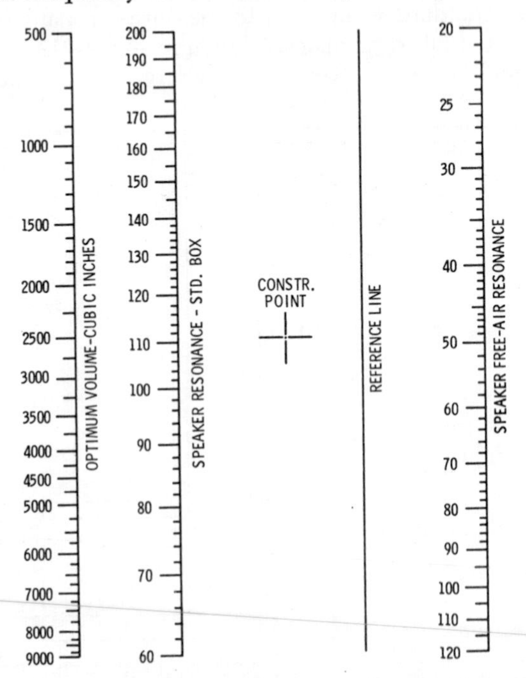

Fig. 4-23. Optimum volume nomogram for use with 8-inch loudspeakers. Standard box in this particular case is one-half cu. ft.

"Using the nomogram of Fig. 4-24, draw a straight line between Point A, the speaker free-air resonance and Point B, the speaker resonance with the "standard box." From the intersection of this line and the reference line, Point C, draw another straight line through the construction point until it intersects with the optimum-volume line. The number read (3000 cubic inches) is the proper volume for this loudspeaker.

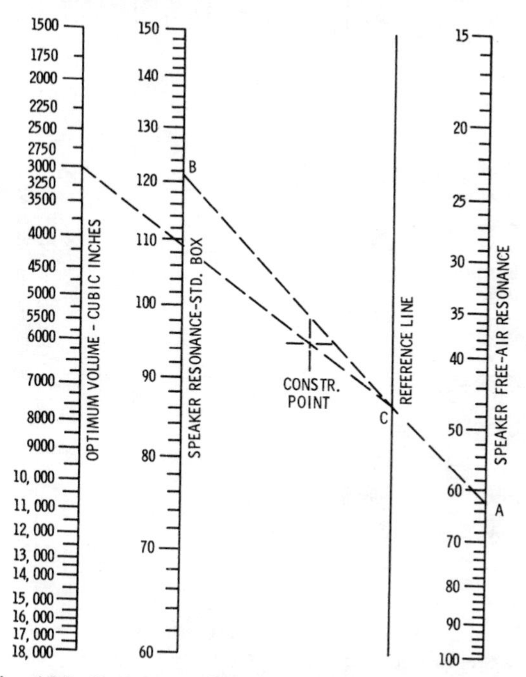

Fig. 4-24. Optimum volume nomogram for use with 12-inch loudspeakers. Standard box here is one cubic ft.

"Unless the reader has reasons for making the enclosure conform to a special shape, the nomogram of Fig. 4-26 can be used for obtaining the proper dimensions for any desired volume. The resulting shape is based on industrial design philosophy that no rectangle will be interesting as an abstract shape until its width equals at least the diagonal of the square on which it is based. The width is, therefore, 1.414 times the height and the height is 1.414 times the depth. Note: The dimensions obtained from the nomogram are *inside* dimensions and must have the material thickness added to them.

"The proper method of using this nomogram is to draw a straight line between the desired volume on the two outside volume scales and obtain the dimensions from the intersection of this line with the three inner scales. The dimensions obtained should be rounded off to convenient numbers such as height (H) = 14⅜″, width (W) = 20⅜″, and depth (D) = 10¼″, for the example cited. The resulting volume will be well within the limits of accuracy required.

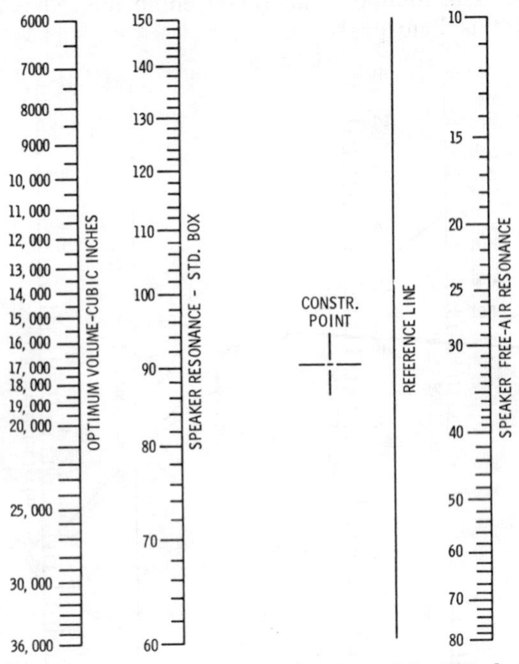

Fig. 4-25. Optimum volume nomogram for use with 15-inch loudspeakers. Standard-box size in this case is two cubic ft.

"The speaker cut-out should be placed toward one end of the enclosure. The table below lists proper size cutouts for 8-, 12-, and 15-inch loudspeakers.

Speaker Size	*Baffle Cut-Out*
8-inch	6¾″ diameter
12-inch	10½″ diameter
15-inch	13¼″ diameter

Tuning & Damping

"The enclosure must now be tuned to the free-air resonant frequency of the speaker. The charts of Figs. 4-27, 4-28, and 4-29 are used for this purpose. They are based on ducts (made of heavy cardboard tubes) which are available for $0.50 each from the Technical Service Department, Jensen Mfg. Div. of the Muter Company, 6601 S. Laramie Ave., Chicago, Illinois, 60638. [*Heavy cardboard mailing tubes, or drawing or blueprint-carrying tubes, available in stationery or drafting stores, may be used.—Ed.*] The proper tube to use will be the largest diameter (inside) which gives a tube length of at least 1½ inches less than the inside depth of the enclosure. In

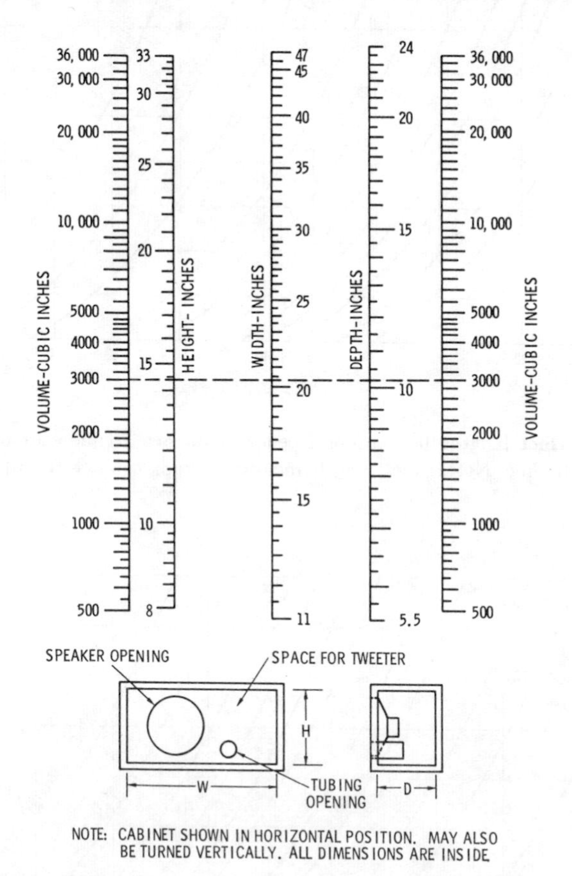

Fig. 4-26. Nomogram used to determine proper enclosure dimensions.

the example chosen above, the duct is 4¾″ i.d. and has a length from the front panel of 3″. (Note: The photos shown at the beginning of this article were taken before the duct was cut to proper length.)

"The speaker and tube should now be installed in the enclosure with the tube being somewhere near the speaker. Although the enclosure volume and tuning are now correct, the system may not

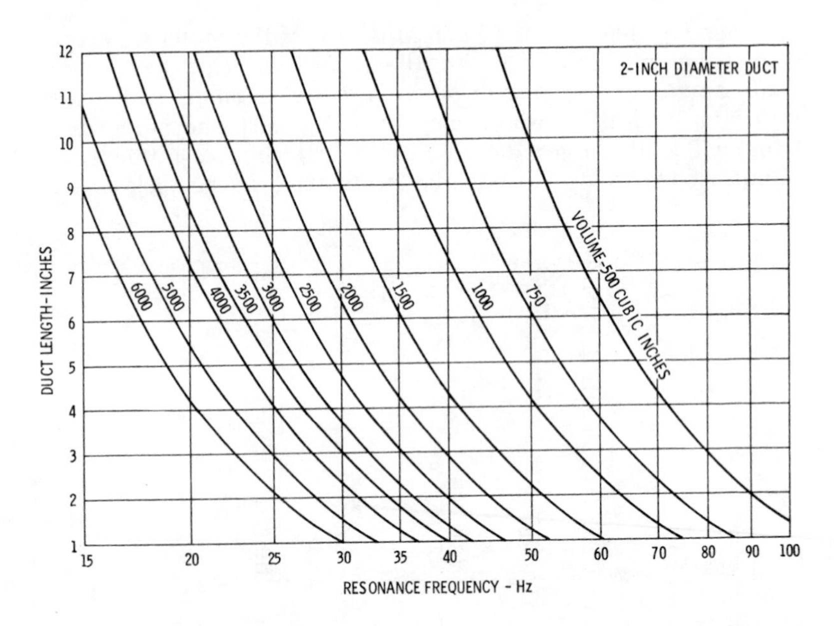

Fig. 4-27. Duct lengths for various free-air resonance frequencies using 2-in. tubing. Note: duct length measured from cabinet front.

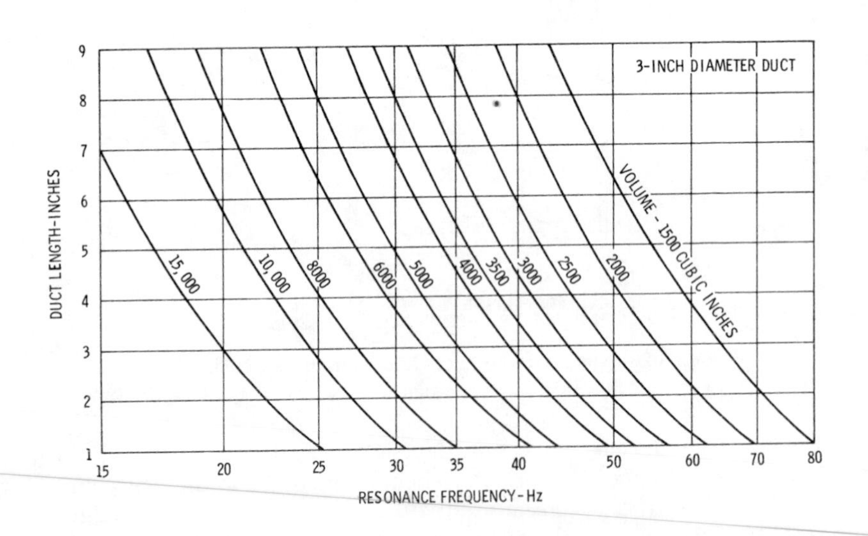

Fig. 4-28. Duct lengths for various free-air resonance frequencies using 3-in. tubing. Duct length is measured from cabinet front.

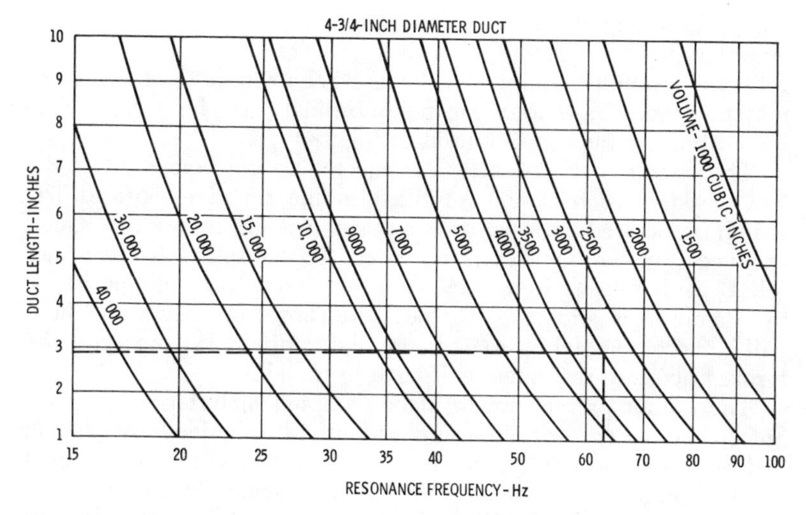

Fig. 4-29. Duct lengths for various free-air resonance frequencies using 4¾" tubing. Duct length is measured from cabinet front.

be free from hangover or boom. The usual method of determining if adequate damping exists by measuring the height of the impedance peaks with the circuit of Fig. 4-21A can often be misleading. A speaker system that appears underdamped with this measurement may be adequately damped when operated with a high-fidelity amplifier. The reason is that the circuit of Fig. 4-21A does not include damping contributed by the amplifier which can be appreciable. For example, an amplifier with a damping factor of 20 appears as a 0.4-ohm resistor in shunt across the voice-coil terminals of an 8-ohm speaker. Many of the transistor amplifiers have damping factors of 50 and greater. The shunt resistance will be less than 0.2 ohm in these cases.

"Damping should be investigated with the circuit of Fig. 4-21B. The value of *R* can be determined if the amplifier damping factor

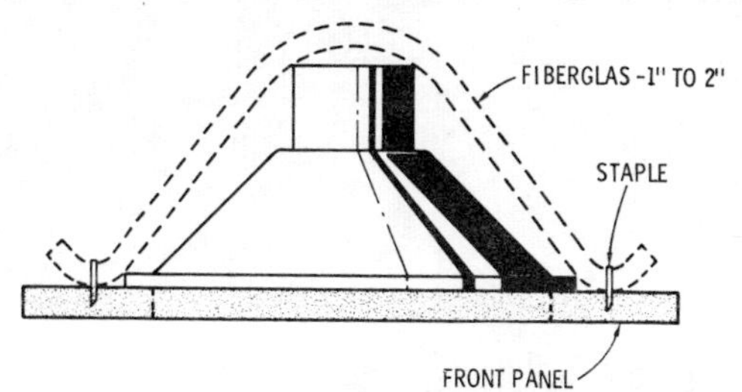

Fig. 4-30. Method of installing *Fiberglas* for damping purposes.

and speaker impedance are known from: R = *speaker impedance/amplifier damping factor.*

"If the amplifier damping factor is not known, a ½-ohm resistor may be used. The battery can be an ordinary flashlight type while the switch can be a push-button or toggle type.

"The circuit of Fig. 4-21B is connected to the voice coil of the loudspeaker which is now installed in the tuned enclosure. The switch is operated between its two positions and the resulting sound produced by the speaker is observed. If the sound is a distinct "click" with no low-frequency boom or "bong" in both positions, the damping is adequate. Chances are, however, that the "click" will be accompanied by some boom and additional damping in the form of acoustic resistance will have to be added.

"The author prefers not to place damping material in the port. Somewhat better over-all results are usually obtained by placing it directly behind the speaker where it can then affect both "tuned circuits." Fig. 4-30 shows the method used. Generally a 1- to 2-inch thickness of lightweight *Fiberglas* (½-lbs. density) stapled around the speaker so that the entire speaker is covered will produce a boom-free click."

SUMMARY

To this point we have discussed how to maximize the radiations from the front and the rear of the cone speaker. Simple box construction has characterized the enclosures discussed, and, in each case, home construction has been feasible. Now, more complex solutions to the problems of converting mechanical motion to acoustical energy with far greater efficiency will be discussed. To accomplish this we now turn to the horn projector.

5

Horn Enclosures

You say you are not the compromising type. You have willingly paid the price (and borne the discomfort) of the 7-liter Ford GT Mark II that won at LeMans last year because performance is the most important single factor you consider. When you hunt you brace yourself to survive the blast of your Weatherby Mark V .387 Magnum. Anything smaller than your 8 × 10 Linhof is a miniature camera to you. Your Aero-Jet Commander shuttles you between your hand-built Norwegian motor sailor in Antigua and your Beacon Hill apartment in Boston where you record the live f-m stereo broadcasts of the Boston Symphony Orchestra on your Ampex MR-70. In short, you demand the very best, and you are able and willing and eager to build your own low-frequency exponential horn. Its very enormity attracts you. A straight, exponential horn for use down to 30 Hz will exceed 16 feet in length, and terminate in a mouth with the man-size dimensions of 11 ft × 11 ft (see Fig. 5-1).

The mouth area is such that its diameter (if a circle) would be about 12¼ feet. The throat area will equal 75 square inches or, if a circle, will have a diameter of approximately 10 inches. All this, of course, will have to be rigid, nonresonant, and open into an area allowing development of 37-foot wavelengths (see Figs. 5-2 and 5-3).

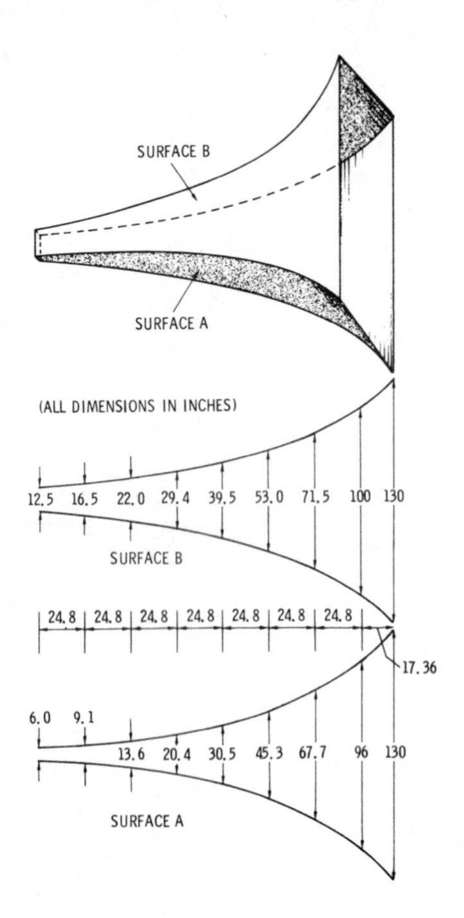

Fig. 5-1. Massive dimensions of a straight horn required to reproduce low frequencies.

TYPES OF HORNS

There are hyperbolic, conical, exponential, catenary, and parabolic acoustical horn expansions (see Fig. 5-4). Three of these are commonly used in making acoustical horns—the hyperbolic, conical, and exponential. The catenary was once offered in a unit called the "Lee Catenary Horn," but it is no longer available. The parabolic horn had its primary use in radio-frequency and microwave devices. Of the three types in acoustical use, the choice is based on the following variables.

Hyperbolic

For horns of the same f_c (low-frequency cutoff) the hyperbolic is the quickest to reach a desirable throat impedance; hence, it is the most efficient down to f_c (see Fig. 5-5). At first glance, this would

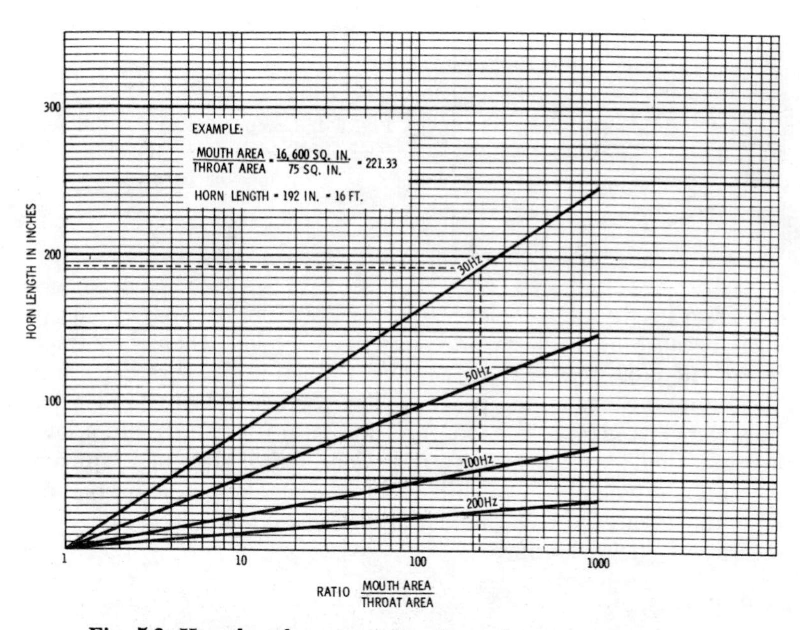

Fig. 5-2. Horn length versus ratio of mouth area to throat area.

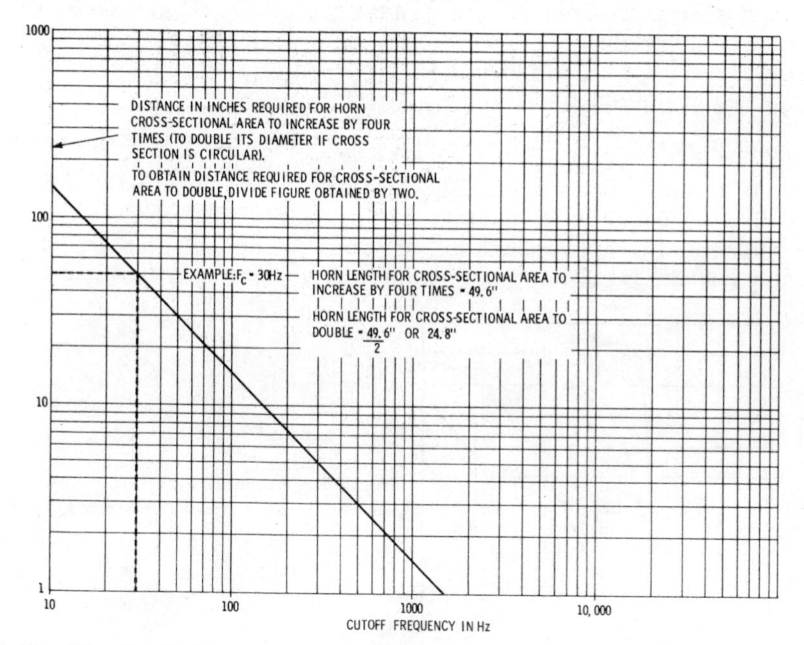

Fig. 5-3. Cutoff frequency versus distance required for cross-sectional area to quadruple.

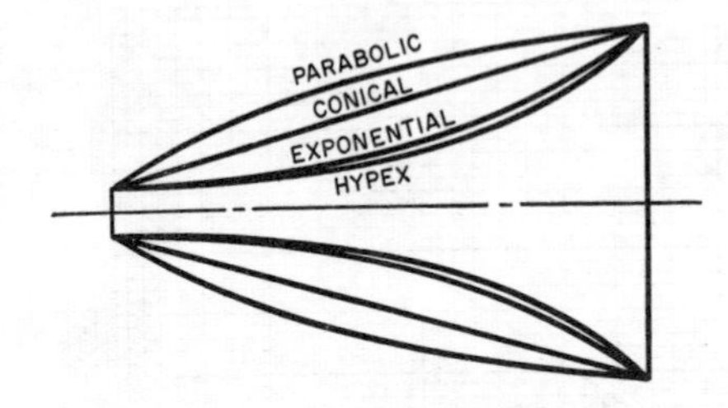

Fig. 5-4. Comparison of the different rates of flare used in the design of speaker horns.

appear to be the best choice. Unfortunately, nonlinearity occurs in horns because of the high pressures that develop when the expansion of the cross-sectional area is slow. When a driver radiates into a horn throat it has a high-pressure, low particle-velocity, and when the wave expands into the mouth of the horn it expands into a low-pressure, high particle-velocity radiation. The higher in frequency the horn is driven above f_c while still traveling along a high-pressure path, the greater the harmonic distortion generated by the nonlinearity of air in the horn.

Conical

Acoustically speaking, a conical horn would theoretically increase its cross-sectional area the fastest of the three types commonly used, and would produce the lowest distortion at higher frequencies; but the conical horn starts to fall off in response at 20 times the frequency of the hyperbolic horn.

Exponential

As has been stated throughout this text, good engineering usually dictates a compromise solution. The expansion rate of the exponential horn is rapid enough to keep pressure and distortion acceptably low and yet it does not begin to "roll off" in frequency response

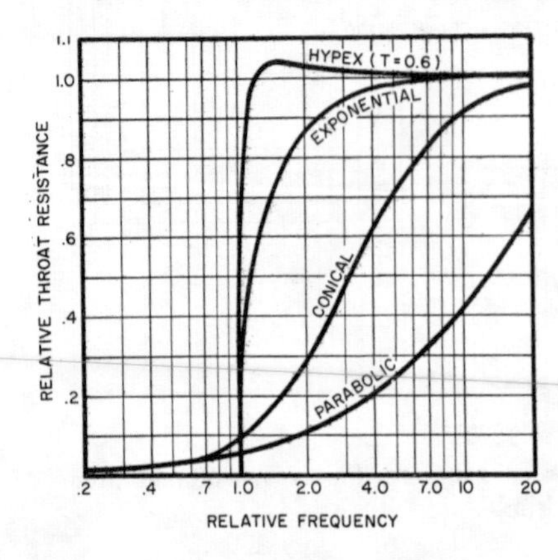

Fig. 5-5. Comparative performance of the horn flare rates illustrated in Fig. 5-4.

until about three times the cutoff frequency of the hyperbolic. For these reasons, most horn designers have chosen the exponential flare progression.

DESIGNING THE WOOFER HORN

To design a woofer horn, several decisions have to be made. First, the lowest frequency to be reproduced should be determined and its wavelength calculated; e.g., 30 Hz = 1100/30 = 36.5. Then, divide the wavelength by three (36.6/3 = 12.2 feet) to ascertain the diameter of the mouth of the horn, if the mouth is in a circular shape.

The mouth need not be circular in the final design and often it is square in shape. Its area must equal, however, the same mouth area, no matter what shape it takes.

Rectangular shapes are hazardous due to the possibility of having standing waves appear across dimensions that would equal submultiples of the wave being produced. This is especially important in working with "folded" horns.

The second consideration is to determine the throat size and area. A large cone driver will be chosen because the free-air cone resonance should be at or below f_c for the horn. To maintain proper efficiency, however, it should be matched to the horn throat through an acoustical low-pass filter of the type seen in Fig. 5-6.

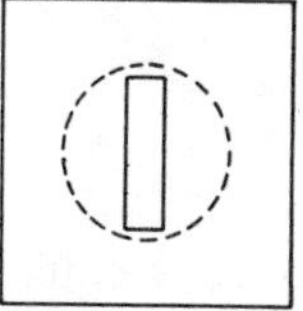

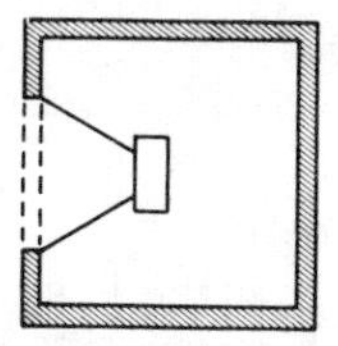

Fig. 5-6. Driver housing employing an acoustical low-pass filter and intended as the source for an exponential horn with a throat the same area as the slot shown.

The size of the slot used in this filter will be determined by the choices of crossover—the higher the crossover frequency, the larger the slot area. The crossover frequency will depend on:

1. The efficiency required.
2. The response of the woofer at higher frequencies.
3. The length of horn that can be tolerated. (A shorter horn would require a larger throat area.)

A good compromise in throat area for a 30-Hz horn using a 15-inch woofer would be 75 square inches, which allows a crossover point of 500 Hz to be safely chosen. The formula for calculating the throat area is:

$$A = \frac{V}{2.9R}$$

where,

A = throat area,
V = volume of air chamber,
R = length of horn within which the horn area doubles.

Thus, for a 15-inch woofer, a rear air chamber of 5400 cubic inches should be built with a slot in front of the cone consisting of 75 square inches in order to have an exponential horn expansion that doubles its cross-sectional area every 24.8 inches. Once the throat and mouth cross-sectional area have been calculated, it can be assumed that the curve that joins them will be exponential.

For practical purposes it is easiest to calculate the distance required to double the cross-sectional area: multiply that length by the number of times it takes to reach an area equal to the specified mouth area. The formula is as follows:

$$\frac{A_2}{A_1} = e^{kx} = 2$$

where,

A_2 = cross-sectional area at the distance x,
A_1 = cross-sectional area at throat,
e = 2.718 (the base of Naperian logarithms),
k = flare constant.

Now, k can be calculated for the chosen low-frequency cutoff (f_c) by:

$$k = \frac{4\pi f_c}{C}$$

where,

f_c = the low-frequency cutoff,
C = velocity of sound in air (13,200 inches per second).

From the formula, $e^{kx} = 2$, a table of Naperian logarithms gives for 2 the value of 0.69315 or roughly 0.7. Thus, for a low-frequency horn with an f_c of 30 Hz, the use of the formula $k = 4\pi f_c/C$ yields $k = 4\pi 30/\ 13{,}200 = 0.028$. And from the formula $e^{kx} = 2$, or $kx = 0.7$, the final calculation, $x = 0.7/.028 = 24.8$ inches, is obtained. Therefore, the cross-sectional area of the horn will double every 24.8 inches of length.

If the f_c of 60 Hz were chosen, then x would equal 12.4 inches, and at an f_c of 120 Hz, x would equal 6.2 inches. Such changes would also make the mouth area proportionally smaller because the f_c would then have a shorter wavelength.

As previously calculated, the mouth area for a 30-Hz horn would be 12.2 feet (146 inches) in diameter and, therefore, it would have a cross-sectional area of 16,900 square inches. The throat would have a cross-sectional area of 75 square inches. Seventy-five square inches would have to be doubled 7.7 times to equal 16,741 square inches. Therefore, the length of the horn would be 7.7 times 24.8 inches, or 16 feet.

Straight horns of this type have been built out of concrete and have been added to a house. Miscalculations tend to be expensive; but results, admittedly, can be very impressive.

FOLDED CORNER HORN

The folded corner horn is an ingenious design calculated to reduce the size of the horn structure by a factor of 8 to 1 by utilizing not only a horn that folds back and forth in its own length while expanding its cross-sectional area, but which also uses the natural corner of a room as the final flare to its mouth.

The folded horn has a long and interesting history starting with the work of early pioneers in electroacoustics such as Kellogg, Stone, Weil, Ephraim, Sandeman, and Horace-Hume. In 1940 this work reached a culmination in a design by Paul W. Klipsch. This design was a folded corner horn of the exponential type that required only one-eighth of the physical space of a straight exponential horn of the same acoustical specifications. This design was quickly dubbed the "Klipschorn" by interested wags, and its many forms are excellent examples of horn design. Figs. 5-7A, 5-7B, and 5-7C illustrate the dual path the horn takes and how it is mated to the corner of a room to provide the final flare of the mouth. The listener actually sits inside the horn. (Fig. 5-8 shows a commercial version of the Klipschorn.)

The Klipschorn is an exponential horn with a nominal cutoff of 40 Hz. It uses a throat area of 87.5 square inches and a back chamber volume of 4700 cubic inches.

The 15-inch woofer is specially designed: it has a 3.2 ohm (d-c resistance) voice coil. However, installed in the horn, its impedance becomes 16 ohms because of the excellent loading and high efficiency of the horn.

One of the disadvantages of the folded corner horn is the impracticality of achieving a phase match at the crossover between the low-frequency and high-frequency horns. To have equal distances from the woofer cone and the high-frequency diaphragm to the listener's ear would require the construction of a tunnel back through the corner that is equal in length to the bass horn. The high-frequency driver would have to be mounted at the rear of this tunnel for correct phasing.

For its size, this type of horn has greater dynamic range and lower distortion at maximum spl than any other type.

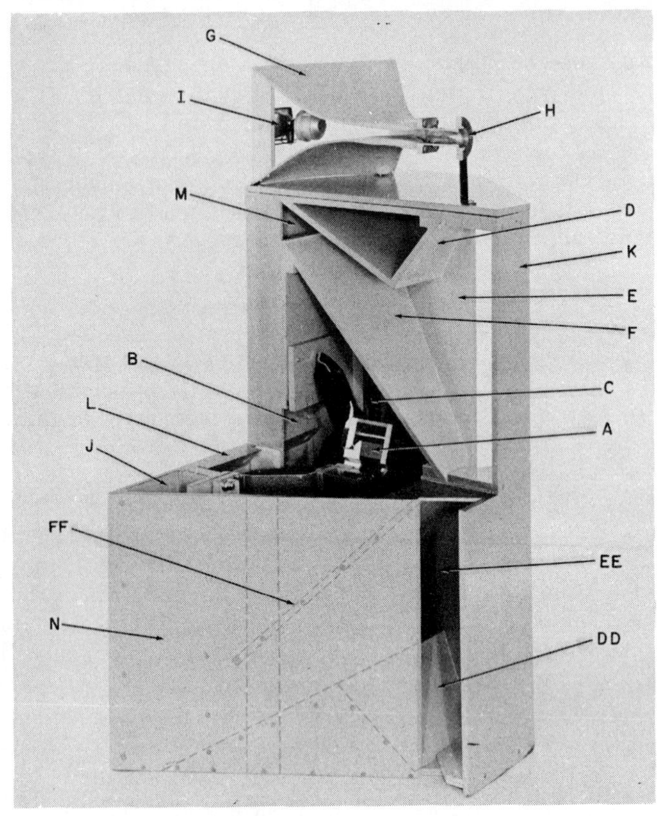

(*A*) *Cutaway view showing midrange and high-frequency speakers mounted on top of the low-frequency horn enclosure.*

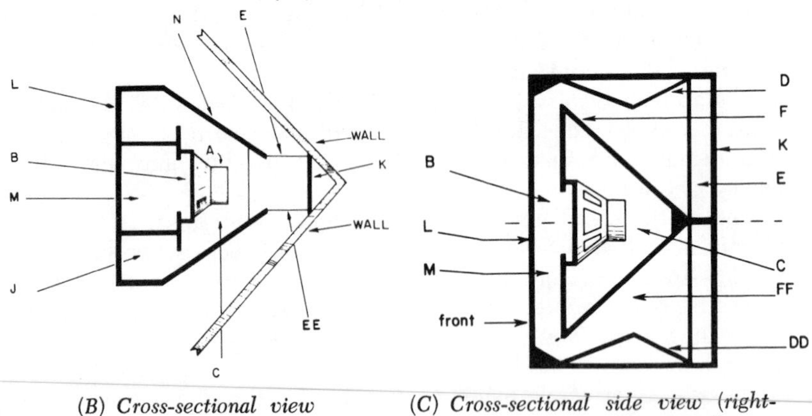

(*B*) *Cross-sectional view looking downward.*

(*C*) *Cross-sectional side view (right-hand side panel removed).*

Courtesy Klipsch and Associates, Inc.

Fig. 5-7. Interior views of a *Klipschorn* enclosure.

Fig. 5-8. The *Klipschorn*—a folded, corner-loaded horn.

Courtesy Klipsch and Associates, Inc.

Fig. 5-9. Large speaker enclosure for theater systems.

Unfortunately, home construction of this complicated device has a long history of unsatisfactory end products. The bass horn, with driver installed, in an unfinished plywood utility cabinet can be purchased at a price well below the cost of "cut-and-try" construction in a home workshop. The constructor is then left with the task of matching the crossover network, high-frequency horn, and high-frequency driver, plus selecting the decorative housing for the units.

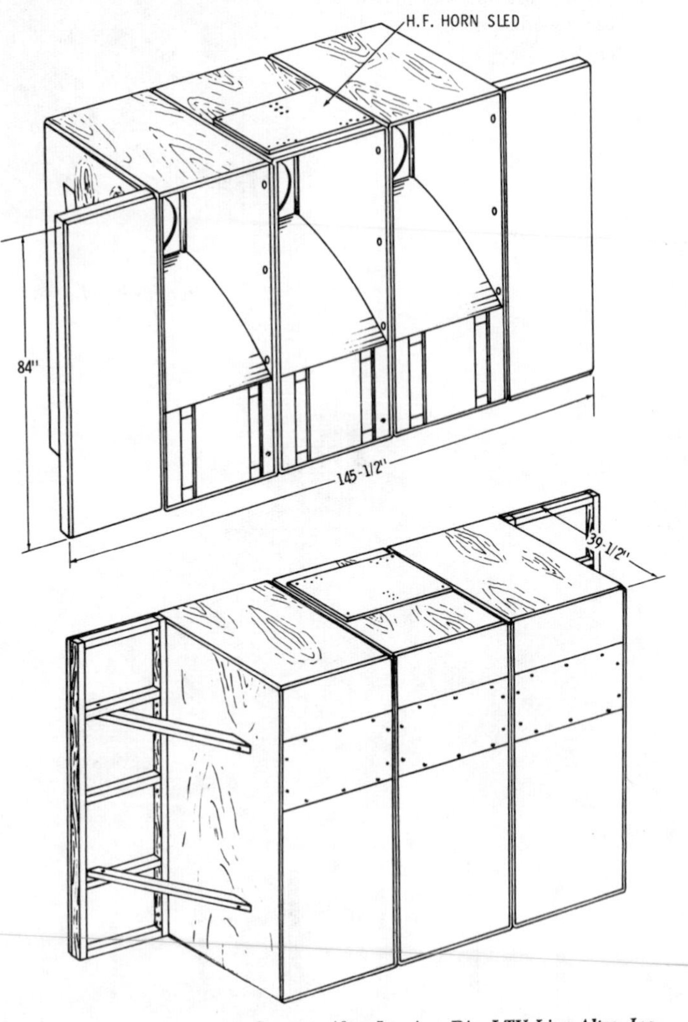

Courtesy Altec Lansing, Div. LTV Ling Altec, Inc.

Fig. 5-10. Dimensions of large horn shown in Fig. 5-9.

ECONOMICAL COMPROMISES

It can be seen from the formulas in this chapter that as the frequency increases, the size of the horn reduces; and the advantage of "horn loading" can economically be enjoyed once the region above 300 to 500 Hz has been reached. (In other words, switch to horn loading at the first crossover point in a lower-cost system.)

Fig. 5-9 shows a commercial version of a straight-horn system with a 70-Hz cutoff. In this commercial theater version, the throat area is greatly expanded at the expense of low-frequency efficiency. Six woofers are used to help raise the bass efficiency below the point where the horn unloads, and the rear of each pair of woofers is coupled to a reflex enclosure to further boost bass efficiency.

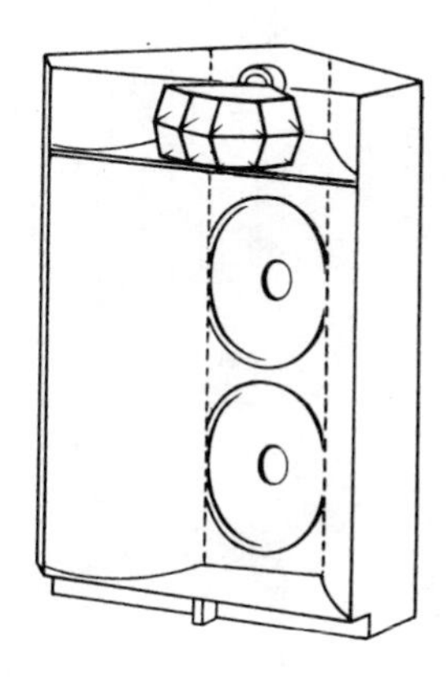

Fig. 5-11. A straight horn with approximately an 80-Hz cutoff.

Perfect phasing of the high-frequency unit with the six woofers is achieved by placing all drivers in the same vertical plane. This unit provides excellent response down to approximately 40 Hz. Fig. 5-10 gives the physical dimensions of the array. Figs. 5-11 and 5-12 illustrate a compromise straight-horn system that enjoyed an exceptional vogue for a number of years. While complex, it is adaptable to home construction if one has had some experience.

A HYPERBOLIC FOLDED HORN

Complete construction details for the Jensen Imperial low-frequency folded horn are provided in this section. Fig. 5-13 shows the Imperial as a free-standing type for corner or sidewall. Fig.

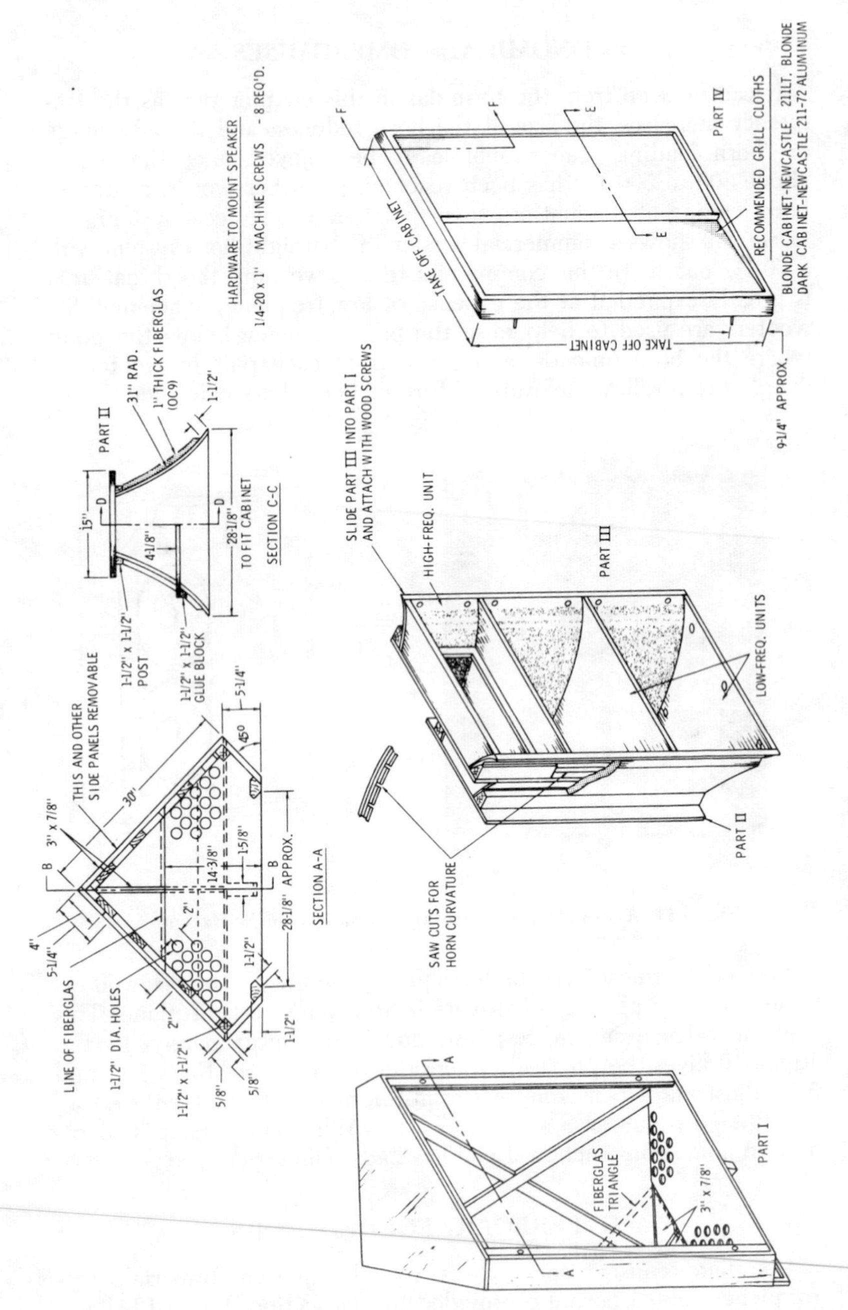

Fig. 5-12. Construction drawings for the

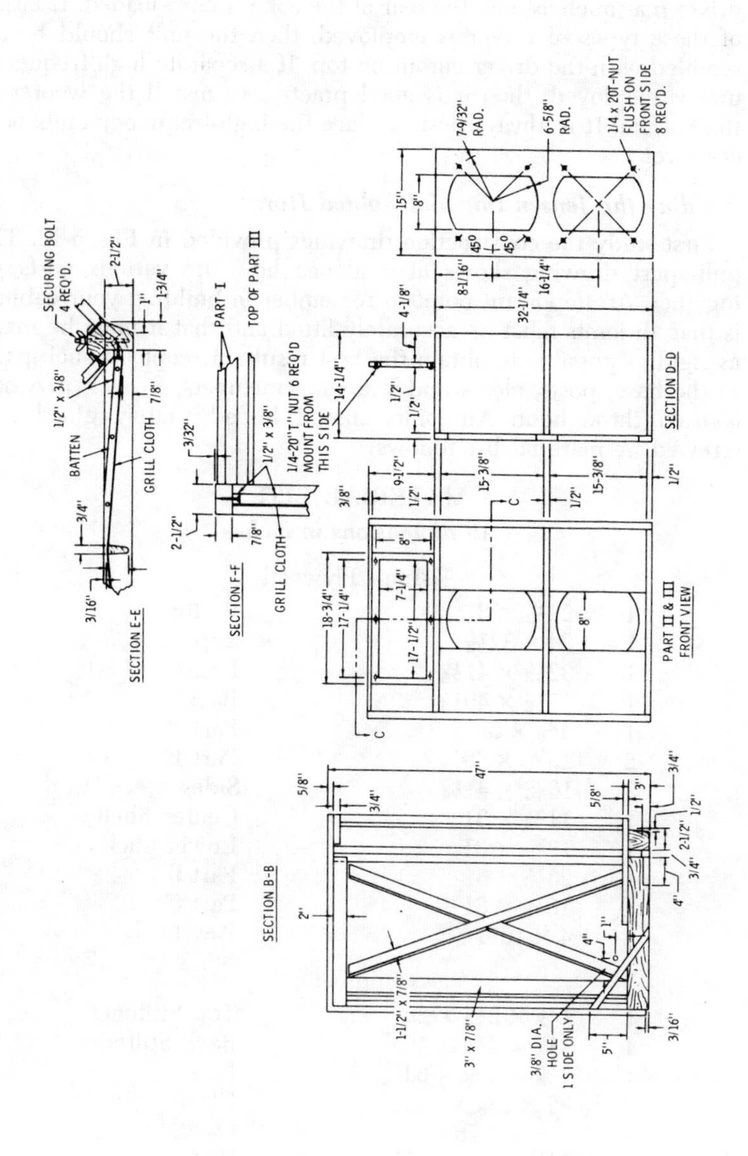

Courtesy Altec Lansing, Div. LTV Ling Altec, Inc.

straight corner horn shown in Fig. 5-11.

5-14 shows the Imperial as a build-in. This is a very fine horn dimensioned for use where a corner is not available and build-in space is. This design allows the use of either a coaxial or triaxial driver inasmuch as only the rear of the cone is horn loaded. If either of these types of driver is employed, then the unit should be assembled with the driver cutout on top. If a separate high-frequency unit is employed, then it is good practice to install the woofer at the bottom. It is always best to place the high-frequency units near ear level.

Building the Jensen Imperial Folded Horn[*]

First study the construction drawings provided in Fig. 5-13. The pull-apart drawing shows at a glance how the various parts go together. An important point to remember in building your cabinet is that all joints must be accurately fitted and that it must be made as rigid as possible to obtain the best results. Except for such parts as the base, posts, cleats and stiffening members, ¾-inch plywood is used throughout. All joints should be adequately glued and screwed. A material list follows:

MATERIAL LIST

(*All dimensions in inches*)

	¾-Inch Plywood	
1	22½ × 31	Bottom
1	24 × 32½	Top
1	32½ × 41¼	Front
1	17½ × 49¼	Back
1	4⅜ × 31	Part A
2	10⁵⁄₁₆ × 49¼	Part E
2	16¾ × 41¼	Sides
1	11¾ × 31	Center Shelf
1	12⁵⁄₁₆ × 31	Lower Shelf
1	23⅛ × 31	Part B
1	15½ × 31	Part C
1	4 × 12¼	Part D
	Lumber	
1	1½ × 2½ × 29	Top Stiffener
1	1½ × 2½ × 40	Back Stiffener
1	1⅝ × 3⅝ × 64½	Base
2	¾ × 2⅝ × 48½	Cleats
1	1 × 1 × 76½	Cleats
2	1½ × 1½ × 48½	Posts

[*] Courtesy Jensen Mfg. Div., The Muter Company.

Begin by cutting out the bottom for the cabinet. The top-view and side-view drawings give the dimensions to follow in laying it out. You'll note that the bottom of the cabinet is ¾ inch smaller all around than the top to let the sides of the cabinet overlap the edges of the bottom. As with all parts, it is important that the bottom be cut squarely since it must fit the sides and front tightly. Next cut the front panel for the cabinet. This measures 32½ inches wide and 41¼ inches high. As indicated in the top-view drawing, the front overlaps the edges of the sides. The 13¼-inch speaker opening is centered in the front panel and on a line 11¾ inches up from the bottom edge. Use a compass and keyhole saw to cut the speaker opening. Each front corner of the cabinet has a 1½-inch-square post 48½ inches long. Detail A shows how each post is grooved on two faces to receive the notched ends of Part A. The grooves are cut ¼ inch deep and ¾ inch wide at a point 8 inches down from the top of the post. These are easily cut by machine with a dado head or by hand with a saw and chisel.

The bottom and front of the cabinet now can be joined together, using glue and screws or nails, and watching to see that the bottom sets in ¾ inch on each side and flush with the bottom edge. This joint should fit tightly at all points. Brace the two parts temporarily to hold them at right angles. Like the bottom, each post is placed ¾ inch in at the corners and fastened to the front panel with flat-head wood screws from the inside. This should bring the grooves in the posts even with the top edge of the front. Part A is made next. This is 4⅜ inches wide and 31 inches long and has a 1¼-inch notch at each front corner to enter the grooves in the posts for a distance of ¼ inch. The rear edge of part A is beveled approximately 20 degrees. Now apply glue to the grooves in both posts and along the front edge of the piece, and fit it in place. The joint across the front should fit tightly like the others. Screws can be used to draw it up tight.

The network and speaker compartments are installed next. The center shelf is made 11¾ inches wide and 31 inches long. The rear edge is beveled 20 degrees to match the beveled edge of part A, and the two front corners are notched to fit around the 1½-inch posts. Part B measures 23⅛ inches long and 31 inches wide and has a 12-square-inch opening cut in the center. Both top and bottom edges of the piece are beveled as shown.

Now lay the cabinet assembly face down. The notched center shelf is attached to the front panel with a 1-inch-square cleat which is cut to fit between the posts and glued and screwed at a point 20¾ inches down from the top edge of the front panel. The notched shelf is glued and screwed in turn to the cleat and posts, after which part B is glued and screwed to the beveled edges of part A and the shelf.

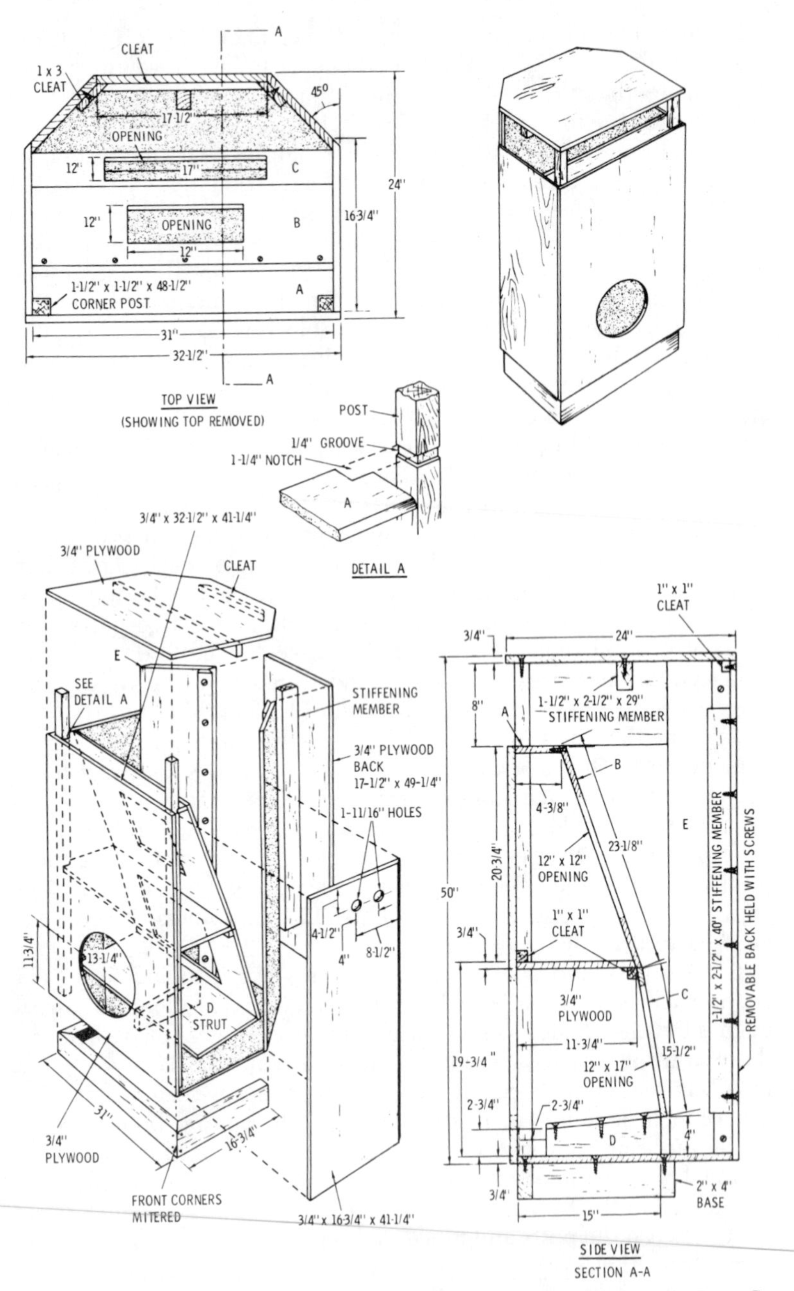

Courtesy Jensen Mfg. Div., The Muter Co.

Fig. 5-13. Construction drawings for the "Imperial" folded horn.

At this point it is best to add a side to the cabinet to give support to the bottom. Each side measures the same, 16¾ × 41¼ inches, and is beveled 45 degrees along the rear edge to be even with part E (see top view). Note in the case of the right-hand side that two holes are bored in it near the top for the H-F and M-F Balance Controls. Glue is applied to all surfaces that the side panel will touch and screws or finishing nails are used to fasten it securely.

The lower speaker compartment is added next. Looking at the sectional view you will see that the lower shelf, which is 12⁵⁄₁₆ inches wide and 31 inches long, is held at a slant by center support D. This part is 4 inches high at the rear and 2¾ inches high at the front and is glued and screwed to the bottom of the cabinet 2¾ inches in from the front panel. The shelf in turn is butted against the side of the cabinet and screwed to the top of the support. A slight bevel is necessary (approximately 6 degrees) at the back edge. Nails are driven through the side and into the end of the shelf. The speaker compartment is enclosed with a 15½ × 31-inch panel, part C, which has a 12 × 17-inch access opening cut in the center. The lower edge is glued and screwed to the rear edge of the shelf, while the upper edge is fastened to a 1-inch (approximately 12 degrees) beveled cleat which is first glued and screwed to the underside of the center shelf. With part C in place, turn the cabinet on its side and install the other side panel, first applying glue to all contacting surfaces. Both sides of the cabinet can be completed now by adding the full length panels, part E. Note in the top view that both vertical edges of these panels are beveled 45 degrees and that rabbets for the removable back are formed with 1 × 3-inch cleats. Like the sides, these panels overlap the edges of the bottom.

The top of the cabinet is of the same shape as the bottom except being ¾ inch larger all around. A stiffening member is added to the underside of the top to increase rigidity and reduce vibration. The top is supported by the posts and parts E to which it is fastened with glue and finishing nails. This leaves the removable back of the cabinet which is held with screws. Like the top, it, too, is fitted with a stiffening cleat placed in the center. Finally, a 1-inch-square cleat is fitted across the opening at the top to provide a screwing surface for the removable back panel. The 2 × 4-inch base is optional, although it does improve the looks of the cabinet. This may be added at a time of construction when it is convenient to drive screws down through the bottom. Screws in counter-bored holes in the base also may be used to attach it. While not shown in the sectional view, the access openings to the speaker and network compartments are covered with ¾-inch plywood panels about 2 inches larger in each dimension and which are held with screws. All nail and screw heads can be sunk slightly below the surface and puttied over.

Constructing the Build-in Imperial[*]

Remember in building the enclosure that all joints must be accurately fitted and that it must be made as rigid as possible to give the best results. All joints should be adequately glued and screwed. With the exception of the base, cleats and stiffening members, ¾-inch plywood is used throughout. A material list follows:

MATERIAL LIST

(*All dimensions in inches*)

	¾-Inch Plywood	
1	24 × 36	Bottom
1	25½ × 37½	Top
1	36 × 60¾	Back
2	25½ × 60¾	Sides
1	3¾ × 17⅝	Part A
1	17 × 36	Part B
1	16 × 36	Part C
1	7⅜ × 36	Part D
1	15⅜ × 36	Part E
1	14½ × 36	Part F
1	21 × 36	Part G
1	17½ × 36	Part H
1	6¼ × 8⅞	Part J
2	8½ × 19⅜	Parts L
1	19 × 19⅜	Speaker Baffle
	Lumber	
1	1½ × 2½ × 51	Part K and Back Brace
2	1 × 3 × 17⅛	Baffle Cleats
1	2 × 2 × 108	Front Stiffener
1	1 × 1 × 204	Cleats
1	2 × 4 × 84	Base

Begin by laying out the bottom. In looking at the side-view drawing, you'll see that the back and front of the cabinet lap the edges at the bottom. This is true of the sides, also. The bottom measures 24 inches wide and 36 inches long. As with all parts, it is important that the bottom be cut squarely, since it must fit the sides, front, and back members tightly. After the bottom is cut to size, you can add the base. This consists of three pieces of 2 × 4-inch material, two being cut 24 inches long and the other 36 inches long. These

---

[*] Courtesy Jensen Mfg. Div., The Muter Company.

are mitered, glued, and nailed together at the front corners and then attached to the bottom of the cabinet with screws driven down through the top. After this, a 1 × 1-inch cleat is fitted along both the front and rear edges of the bottom, keeping each cleat flush with the edge. These cleats provide additional support for the back panel and the front members of the cabinet.

The back of the cabinet is made next. This is merely a panel 36 inches wide and 60¾ inches long which is glued and nailed securely to the rear edge of the bottom. Check the two members with a square to see that they are at right angles and then brace temporarily to hold them so. You'll notice that a 1½ × 2½ × 39-inch stiffening member is added to the center of the back, on the inside, 5 inches up from the bottom. Before this is attached with glue and screws, a ¾ × 2½-inch notch is cut in the front edge 9⅞ inches down from the top. A 1 × 1-inch cleat fitted across the top on the inside completes the back.

The sides of the cabinet are cut 25½ inches wide and 60¾ inches high. At this stage, one side panel is added to help strengthen the assembly made thus far. Glue and nail this adequately to the edges of both the back and the bottom.

The compartments can be pre-assembled and then installed as a separate unit, or they can be built up, piece by piece, within the cabinet. All the pieces in the speaker compartments are made exactly 36 inches long, the width of the bottom. The side-view drawing gives the widths of the various parts which are keyed to the material list for reference. Start with part A. This is a center strut that supports part B and is glued and screwed securely to the bottom of the cabinet, 2⅜ inches in from the front edge. Part B is 17 inches wide and 36 inches long and is beveled 7 degrees along the rear edge. Install part B by applying glue and driving nails down into the strut and in through the sides of the cabinet. Cut part C next. This is cut 16 inches wide and 36 inches long and is beveled 35 degrees along the rear edge. A 1 × 1-inch cleat is fastened to the underside along both the front and rear edges, the rear cleat being beveled 35 degrees to match the bevel on part C. Part C is installed 18⅝ inches up from the bottom and set in ¾ inch from the front edge.

Next, cut and add part E. This is 15⅜ inches wide and 36 inches long and is beveled 11½ degrees along the upper edge. It is glued and nailed to the edge of part B and screwed to the rear cleat attached to part C. The compartments are completed by cutting parts F and G. Part F is cut 14½ inches wide and 36 inches long and is beveled 55 degrees along the lower edge and 30 degrees along the upper edge. Part G is cut 21 inches wide and 36 inches long and is beveled 55 degrees along the upper edge and 25 degrees along the lower edge. This leaves part H to be fitted across the back of

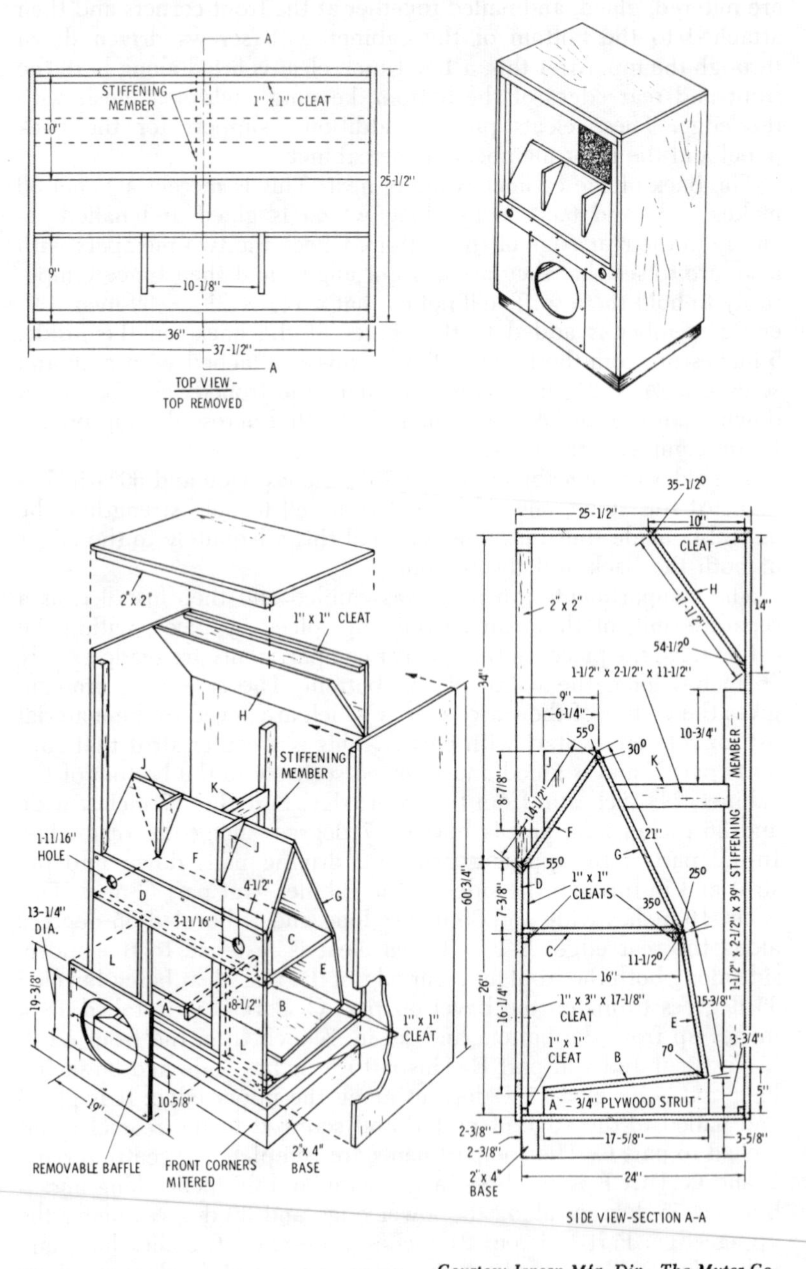

Courtesy Jensen Mfg. Div., The Muter Co.

Fig. 5-14. Build-in version of the "Imperial" folded horn.

the cabinet at the top. This piece is 17½ inches wide and 36 inches long and is beveled 35½ degrees along one edge and 54½ degrees along the other. This is glued and screwed to the back, 14 inches down from the top, and also glued and nailed to the side of the cabinet, driving the nails in through from the outside.

Now place the cabinet on its side to add the other side panel and apply glue to all contacting edges. Nail the side in place from the outside as before. The top measures 25½ inches wide and 37½ inches long and is nailed to the top edges of the back and sides. Parts J and K are added next. Parts J are triangular speaker-mounting blocks, spaced 10⅛ inches apart, and are fastened to part F with screws from the inside. Part K is a brace that fits the notch in the stiffening member at the back and rests against part G to which it is glued and nailed.

The upper part of the cabinet at the front is strengthened by framing the three sides with 2 × 2-inch pieces, driving the screws into the sides and top from the inside. A speaker-baffle opening is framed in the lower part of the cabinet at the front by adding 8½ × 19⅜-inch pieces at each side. These pieces (parts L) are supported at the top by the cleat provided and nailed to the front edge of the bottom. Nails also are driven through the sides of the cabinet. One by three-inch cleats, straddling the inner edges of these pieces, provide rabbeted edges to support the removable speaker baffle. These cleats are 17⅛ inches long and are screwed in place from the inside. The speaker baffle is held with screws only, and the 13¼-inch circular opening is centered at a point 10⅝ inches up from the bottom.

All that remains is cutting and fitting part D. This, like the speaker baffle, is held with screws only and measures 7⅜ inches wide and 36 inches long. A $1\frac{11}{16}$-inch hole is bored at each end for controls and these are centered 4½ inches from the ends and $3\frac{11}{16}$ inches down from the top edge. The ends of part D are supported by cleats which are screwed to the sides of the cabinet. Screws are driven into these cleats, as well as into the edges of parts C and F.

All exposed nail heads and screws can be sunk slightly below the surface and puttied over to conceal them.

SUMMARY

It should be noted that, except in the case of the straight exponential horn of full size and the large Klipschorn, the compromised versions of the horns illustrated can and may be outperformed by a large infinite baffle or a good combination-type system. Horns, when the design goals are left uncompromised, produce results which are uncompromised. Once compromise is accepted, the use of a combination-type enclosure should be considered.

6

Combination Enclosures

The infinite baffle, bass-reflex, and horn enclosures .demonstrate the three basic concepts that can be used to match the cone of a speaker to a room (helping the heavier mass—the cone—meet efficiently a lighter mass—the room air). Much like the three passive circuit elements in electronics—the resistor, the capacitor, and the coil—the infinite baffle, bass-reflex, and horn can be combined to form "circuits" that will accomplish what the one element cannot do alone. Actually, the horn driver is housed in a very small, infinite-baffle enclosure (the rear wave of the cone never meets the front wave), and the front of the cone is then fed to the throat of the horn.

COMBINATION BASS-REFLEX AND HORN ENCLOSURES

One of the disadvantages of a horn is its mammoth size if it is not folded. When it is folded it becomes awkward to maintain proper phase relations between the low-frequency and high-frequency units. When sound first came to motion pictures, full-length horns were often used behind the screens. Both size and phasing problems came rapidly to the fore, and some of the finest audio engineering talent ever assembled in a single industry was brought to bear in obtaining a solution.

The design that finally predominated (so much so that this type is used more often professionally than all other types put together) was a combination bass-reflex and straight horn. A horn with a large throat area, designed with a quite rapid flare rate and with an f_c just below the 100-Hz region to keep its length short, was combined with a tuned bass-reflex enclosure for the back of the cone. This resulted in a "stepped" response curve of the type shown in Fig. 6-1.

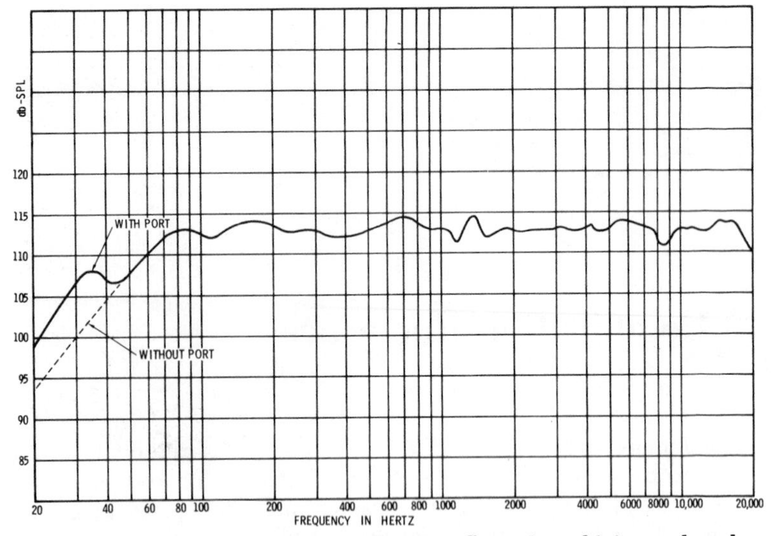

Fig. 6-1. Frequency response curve showing effect of combining a short horn with a bass-reflex enclosure.

If this enclosure is listened to at a normal room distance, the output from both the port and the horn will be heard. In large theaters, the horn projection is the main output heard at the greater distances involved in the larger volume rooms. Thus, as the volume of a space increases and its low-frequency reverberation increases as well, this type of enclosure "matches" the room size with reduced low-frequency acoustical output at a distance, while maintaining full response at shorter distances or in smaller rooms.

The Altec Lansing "A-7" (shown in Fig. 6-2) is a commercial example of this type of combination. Fig. 6-3 illustrates the same functional unit housed in a furniture-style outer cabinet. Fig. 6-4 gives the dimensional details necessary for successful construction of the enclosure.

As mentioned above, a horn can be considered as an acoustical extension to a driver housed in an infinite baffle. Just a jump away in thinking is the idea of attaching a horn to the port of a bass-reflex enclosure. Fig. 6-5 illustrates the basic principle of the arrangement.

Fig. 6-6 shows a version of this type which utilizes the corner of a room as the mouth of the horn. The long slot down the rear of the cabinet serves as a modified port and the front of the driver

Courtesy Altec Lansing,
Div. LTV Ling Altec, Inc.

Fig. 6-2. Combination horn and bass-reflex enclosure.

Courtesy Altec Lansing,
Div. LTV Ling Altec, Inc.

Fig. 6-3. The system of Fig. 6-2 in a furniture cabinet.

continues to feed directly to the room. (See Fig. 6-7 for construction details.)

RESONANT AIR COUPLERS

In the very early 1950's, bass couplers became of interest to the owners of already huge sound systems. They built (usually under the floor of the living room) separate, enormous, coffin-size air couplers tuned to resonate anywhere from 20 Hz to 30 Hz. Special low-frequency crossover networks were built, and the owners of large horn and bass-reflex systems eagerly searched for recordings that contained a note low enough to sound the coupler. These systems were, in essence, either a large horn or bass-reflex combined with a large but narrowly tuned resonant air column.

An interesting set of combinations appeared on the market in the mid-1950's in an enclosure called the *Karlson* (see Fig. 6-8). It

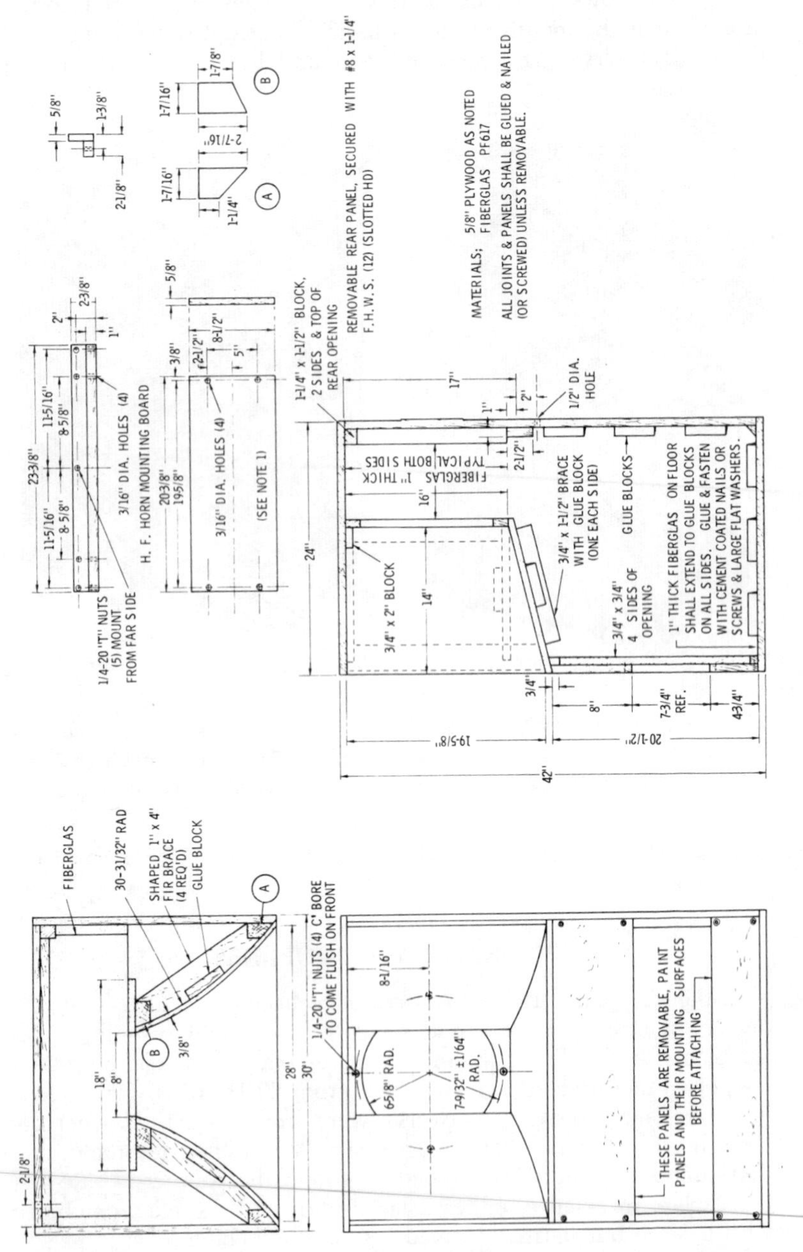

Couresy Altec Lansing, Div. LTV Ling Altec, Inc.

Fig. 6-4. Construction drawings for the enclosure shown in Figs. 6-2 and 6-3.

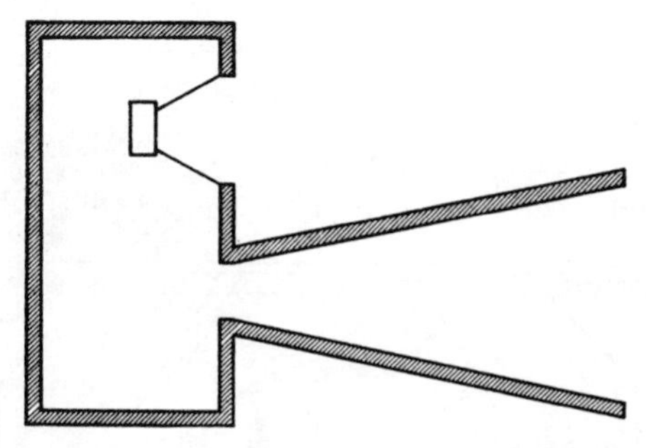

Fig. 6-5. Illustration of the concept of a horn-loaded port on a bass-reflex enclosure. This is usually accomplished by rear loading the driver through a slot at the rear of the enclosure.

contained elements of the horn, infinite baffle, bass-reflex, and a resonant coupler. (See Fig. 6-9 for the designer's view of the combinations achieved.)

Care must be taken in designing combinations to ensure that they do not actually conflict with one another.

ACOUSTIC LABYRINTH

The acoustic labyrinth (such as the one shown in Fig. 6-10) is the only type of enclosure that can lower the low-frequency resonance of a driver.

Fig. 6-6. A corner-type enclosure with a horn-loaded port. The horn is formed by the corners of the room and the sides of the enclosure.

Courtesy Klipsch and Associates, Inc.

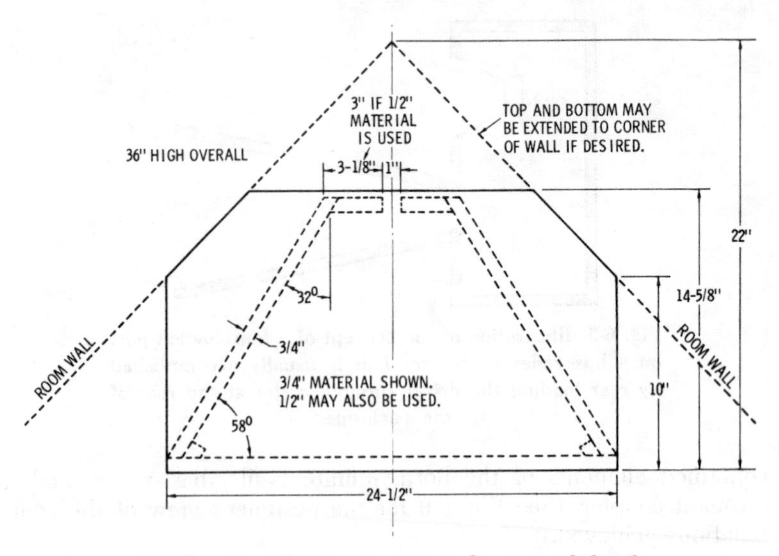

(A) Top view showing important dimensional details.

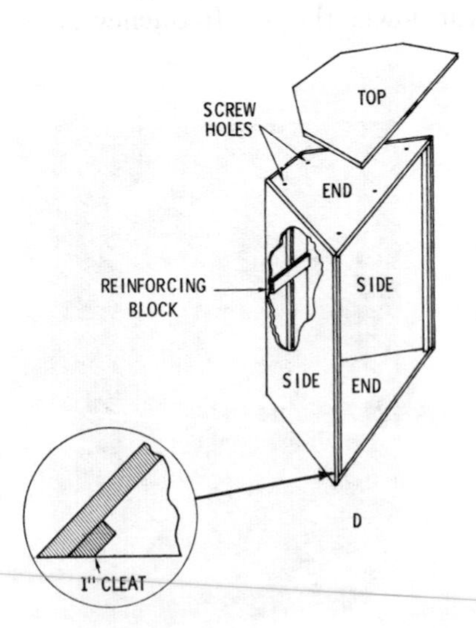

(D) Glue and nail the narrow 1-inch beveled cleats to inside front of sides. Beveled edges should be flush with bevel of sides as shown in detail cross section.

Glue and nail reinforcing block across slot in inside of housing. Tack plastic screen over slot on inside.

Fit top to end and drill end with screw holes as shown, allowing ⅝-inch overhang for ½-inch material, ⅞-inch overhang for ¾-inch material. After fitting, remove top for sanding and finishing. Cement a 1-inch length of sponge-rubber tape beside or over each screw hole. Then attach top with five, short, No. 8 wood screws. This should be done from inside of housing, using screw holes already drilled. Fasten metal furniture gliders to bottom, one at each corner.

Fig. 6-7. Construction details for the

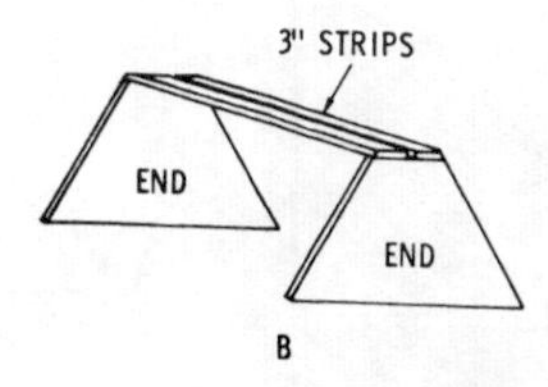

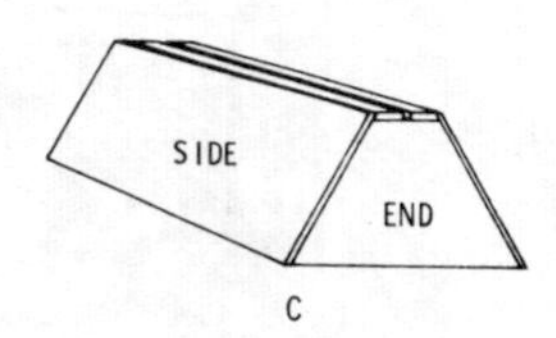

(B) Fasten 3-inch strips to ends, placing beveled edges of strips flush with edges of ends. This leaves slot between non-beveled edges of strips. Glue and nail with finishing nails.

(C) Attach sides to ends. Make sure beveled front edges of sides are flush with ends. Glue and nail, spacing nails about two inches apart.

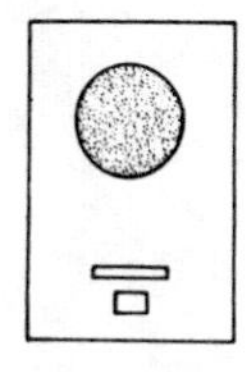

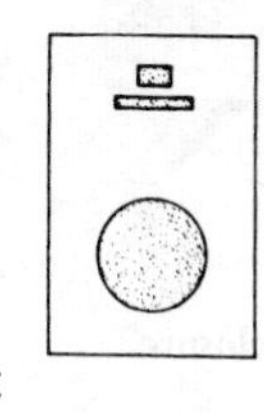

E

(E) Fit speaker drive units to front panel, but do not mount. If bolts are to be used, drill necessary bolt holes.

When single or coaxial speaker is used, fit front panel with 10 ½-inch circular hole at top. This hole will accommodate 12-inch, 15-inch or even 18-inch cone speakers.

When mid-range or high-frequency drivers are used, fit front panel with small holes at the top so that speakers will be at the correct listening level.

Cut out cardboard gaskets for all speaker holes making sure the gasket on the bass driver is sufficiently thick to prevent the cone from striking wood. Add extra gaskets if necessary and cement in place before mounting speakers.

After all sanding and finishing is completed, mount the speakers using screws or bolts. Crossover networks or elements may be mounted on inside of front panel or inside bottom end. Wire leads may enter housing through slot, or a terminal strip may be attached to the side, with through screws serving as connectors.

If grill cloth is to be used inside opening, tack on inside of front panel before attaching speakers. Grill cloth should be of soft, open-mesh material to prevent rattling, flapping, or excessive absorption of sound waves.

Attach sponge-rubber tape around front edge of housing as a cushion between it and front panel. Use rubber cement.

Front panel may now be screwed to housing. Skew the housing slightly, if necessary, to line it up properly.

Courtesy Klipsch and Associates, Inc.

corner-type enclosure shown in Fig. 6-6.

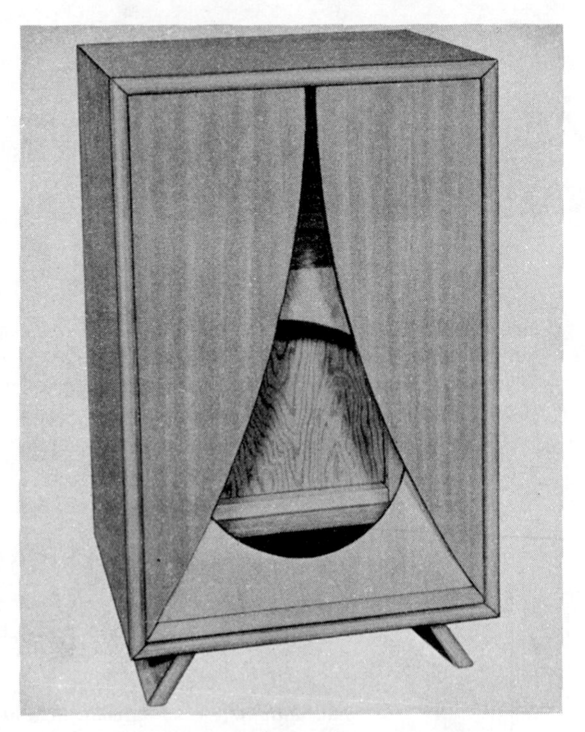

Fig. 6-8. *Karlson* speaker enclosure.

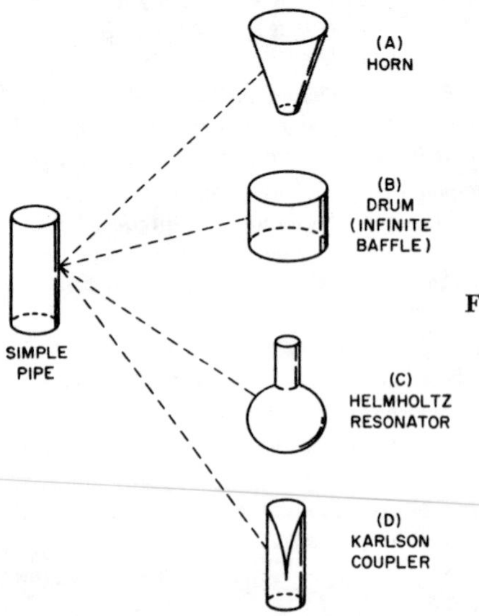

Fig. 6-9. Origin of the *Karlson* speaker enclosure.

An acoustical labyrinth is a bass-reflex enclosure combined with a long, folded, acoustically damped pipe. The driver is considered properly matched to the labyrinth when the low-frequency resonance of the driver, expressed as a wavelength, is equal to four times the length of the labyrinth. For example, a driver with a resonance of 40 Hz should be used with a labyrinth that is seven feet long, since the wavelength of the resonant frequency equals the velocity of sound divided by the resonant frequency: 1100/40 = 28. To find the quarter-wavelength divide the resonant wavelength by four: 28/4 = 7 feet.

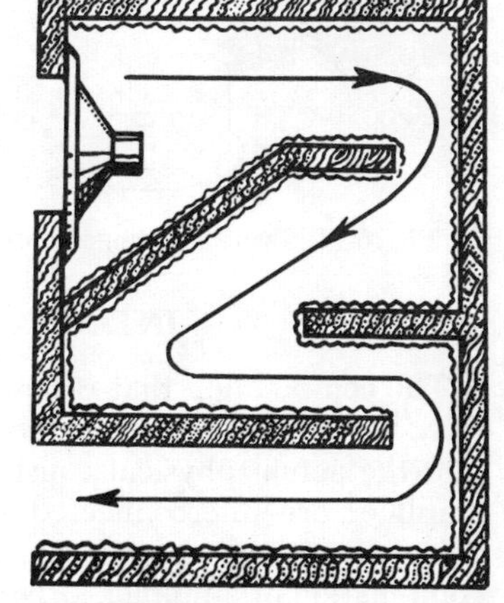

Fig. 6-10. ***Stromberg-Carlson*** **acoustic labyrinth.**

If the driver has a 50-Hz resonant frequency and is placed in a labyrinth seven feet in length, the resonance of the driver will be lowered. While this might seem to be an easier way to get bass, the lowered resonance is at the expense of linear frequency response. Labyrinths are large and phasing can be a problem.

RESONANT COLUMNS

The current high-fidelity market has seen a rash of resonant columns. (These should not be confused with the in-line columns used in public-address work.) The resonant column can be described as a cross between a direct radiator (infinite baffle) and a coupler (a tuned pipe). The vent in the pipe or tube is placed at some distance from the driver, which is usually mounted in the top or bottom of the tube.

Polar characteristics of these units are extremely ragged, and when listened to in a large room or at close range, the bass and treble become disassociated and the sound becomes as if it were played on different manuals of an organ at slightly delayed intervals. (Fig. 6-11 illustrates this type of enclosure.)

The "RJ" was a commercial version of the tuned cavity that offered excellent performance in a small size. Lack of efficiency was one of its problems, plus the usual polar response irregularties.

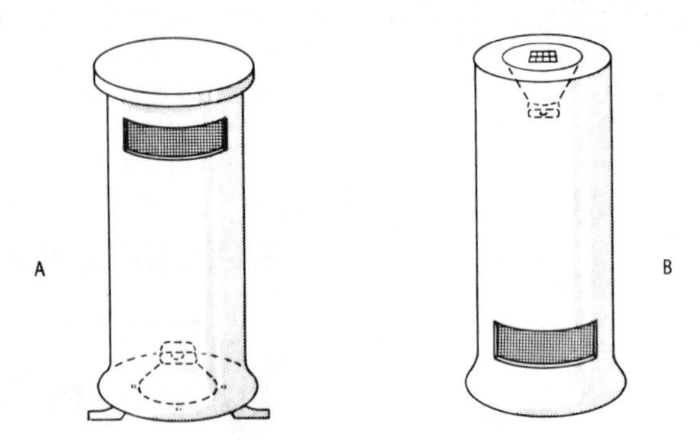

Fig. 6-11. Typical arrangements of resonant-column speaker enclosures.

INTEGRATED SPEAKERS

The combination that is now on the horizon, but has not yet been fully developed, is the integration of a solid-state power amplifier into the actual physical construction of the driver. The driver and amplifier are interconnected so that the feedback path from the voice coil of the speaker can be adjusted to fully match the total combination of amplifier, driver, and enclosure. At the present time, some experimenters are working with small infinite baffles containing a large magnet, large cone drivers with a solid-state amplifier built into the driver frame. Speaker systems of this type can exhibit spectacular transient response because they are very highly damped.

Remembering that the drivers and enclosures are passive devices, as are resistors, capacitors, and coils, it is of interest to note the revolution just beginning in electronics to replace as many passive components with active elements as cost and complexity will allow. No doubt, functional circuit design versus passive circuit design will eventually affect speaker enclosures as well.

SUMMARY

The fundamental methods that can be chosen to help a driver move air either more efficiently or more uniformly have now been discussed. The high-frequency drivers that are not included in the low-frequency enclosure design but have to operate in conjunction with it, must now be matched to the woofer enclosure. To do this we require crossover networks.

7

Crossover Networks

As each type of enclosure was discussed in earlier chapters it was noted that few enclosures allow the driver to operate over an extended frequency range. Usually the low-frequency performance of the enclosure can be improved if the upper frequency of the woofer-driver is restricted. In the case of a horn-loaded driver, a ratio too great between the f_c of the horn and the highest frequency to be reproduced causes increased distortion. This is due to the slow flare that results at the throat of the horn.

HOW TO PICK THE CROSSOVER FREQUENCY

A number of factors enter into the choice of the correct crossover frequency. Ideally it should fall at a frequency low enough to miss the predominant voice frequency range, 500 to 2000 Hz. In a very wide range system of 20 to 20,000 Hz (or 10 octaves), asking each of two drivers to carry five of the octaves results in a crossover around 600 Hz:

20–40 Hz	1st octave
40–80 Hz	2nd octave
80–160 Hz	3rd octave
160–320 Hz	4th octave
320–640 Hz	5th octave
640–1280 Hz	6th octave
1280–2580 Hz	7th octave
2580–5120 Hz	8th octave
5120–10,240 Hz	9th octave
10,240–20,480 Hz	10th octave

Another criterion that can be used is the distribution of power to the speaker system. Fig. 7-1 shows the typical power distribution by frequency of a symphony orchestra. Here it can be seen that the main power demands fall below 500 Hz, indicating that fre-

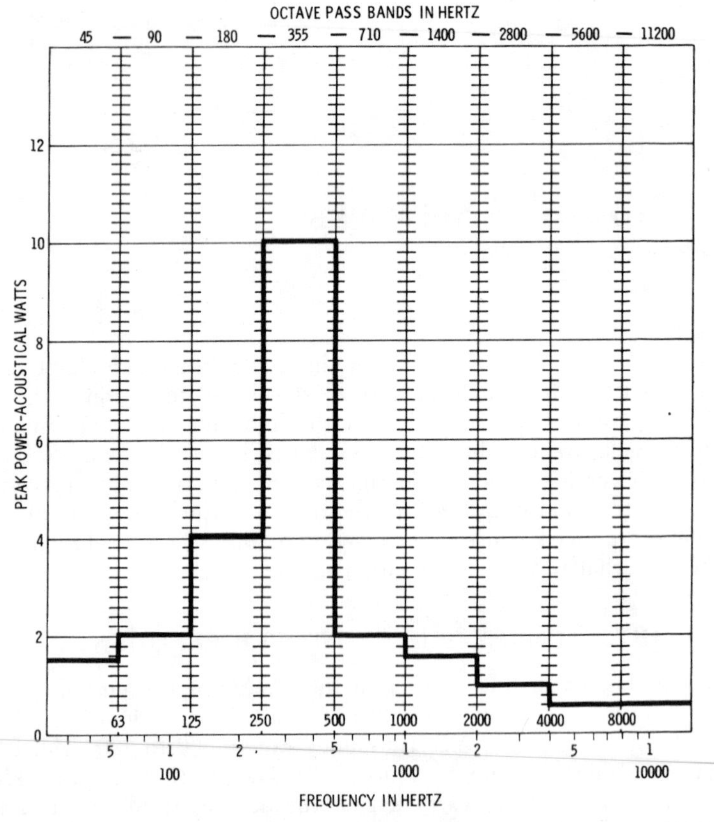

Fig. 7-1. Power distribution in octave bands of a typical orchestral composition.

quency (500 Hz) as a good crossover point inasmuch as woofers can handle more power than high-frequency drivers.

HORN-RATIO EFFECT ON CROSSOVER FREQUENCIES

If a full horn-loaded system is employed, the ratio between f_c and the highest frequency to be reproduced must be kept small or "horn distortion" will become excessive. For example, if 30 Hz is

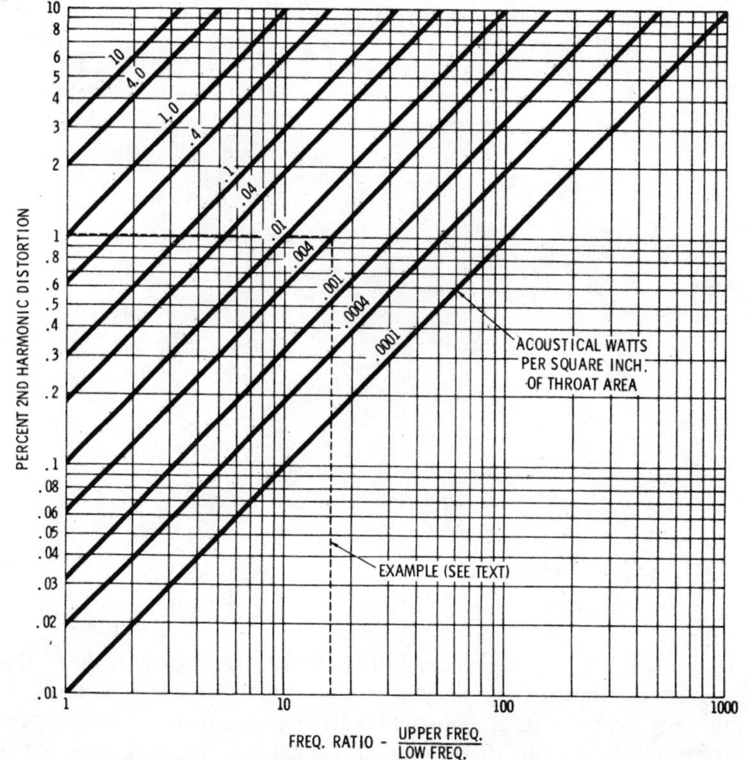

Fig. 7-2. Graph illustrating the effect of the ratio between f_c and the highest frequency to be reproduced, on the amount of second-harmonic distortion in a full horn-loaded system.

the f_c of the horn system, and a 500-Hz crossover point is chosen, the ratio of 500:30 is slightly less than 17 to 1. Referring to Fig. 7-2 it can be seen by following the dotted-line example that 4 milliwatts per square inch can be achieved with 1-percent second-harmonic distortion. With a horn that has a 50 square-inch throat area the total acoustical output, while staying below 1-percent distortion, would be 0.004 × 40 = 0.160, or 0.16 acoustical watt. In a 50-percent efficient unit, an input of 0.32 electrical watt would give an acoutical output of 0.16 watt at 1 percent. If 5-percent distortion were tolerable to the listener, then 0.4 acoustical watts per square inch would be a usable figure; and 50 square inches of throat area would allow 50 × 0.4 or 20 acoustical watts to be generated before 5-percent second-harmonic distortion would be exceeded at the upper frequency. Thus a 40-watt electrical input could be handled with less than 5-percent distortion in the 20-watts acoustical output of the unit that is 50 percent efficient. The distortion in each case is greatest at the higher of the two frequencies.

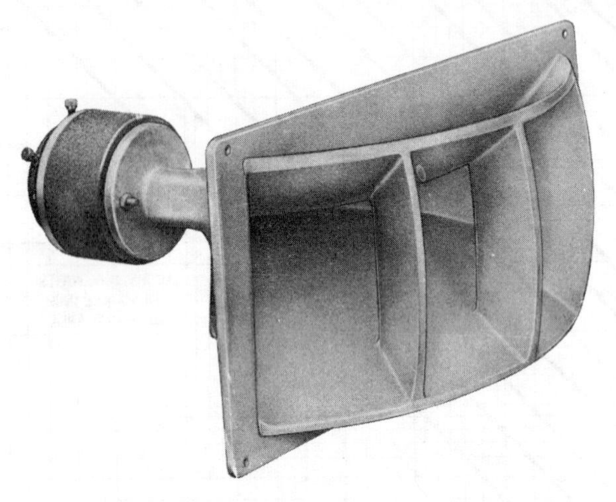

Fig. 7-3. High-frequency driver and horn capable of a frequency ratio of 40 to 1 with low distortion.

If a single horn is used from 500 Hz to 20,000 Hz, the ratio of the upper frequency to f_c is 40 to 1. Thus at the highest frequency it could only produce 0.01 acoustical watts per square inch with 4-percent second-harmonic distortion; at 10,000 Hz it could produce 4 acoustical watts per square inch at 4-percent distortion. Fortunately the power required at higher frequencies is very low, as Fig. 7-1 has already illustrated. If the driver and horn illustrated in Fig. 7-3 were driven to an output of 0.01 acoustical watt (4-percent distortion at 20,000 Hz) it would have an spl at 4 feet of approximately 93 db.

Four-percent distortion of 20,000 Hz (which is the second harmonic of the 10,000-Hz fundamental) has no meaning to the listener's ear, since he cannot hear that high anyway. The highest frequency of interest, as far as distortion is concerned, is one-half the highest frequency one can hear.

These calculations indicate why speaker distortions of 5-percent are considered excellent if achieved at usably high acoustical output levels. All too often speaker distortion figures are quoted at unusably low acoustical output levels; and great caution should be exercised in evaluating distortion figures quoted without distance, sound pressure level, and input power being clearly correlated to the figure given.

CROSSOVER PHASE SHIFTS

Each time a crossover is used, phase shifts occur, with resultant irregularities in the signal as it is reproduced in the two-octave crossover area; therefore, it is important to use as few crossover points as possible (see Fig. 7-4).

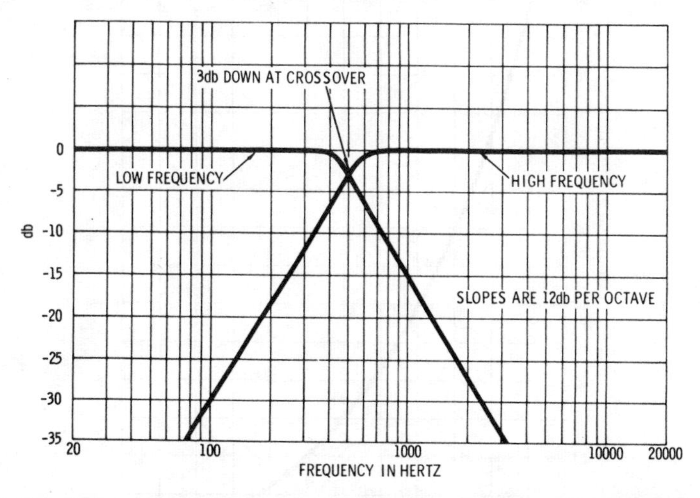

Fig. 7-4. These curves illustrate how each driver remains partially effective through a two-octave range—250 to 500 Hz, and 500 to 1000 Hz.

No crossovers would be the best choice except for the inability of a single driver to cover the entire frequency range smoothly and efficiently, and for *Doppler distortion.* While experimenters and others have built systems with as many as four crossover points, the overwhelming selection of professional designers has been for either one or two, with one crossover point being universally considered the optimum if sufficiently well designed drivers are available.

In three-way systems, 500 and 5000 Hz are almost the only professional choice; and in two-way systems 500 or 800 Hz have found favor, with 500 Hz being considered the best overall compromise.

INSERTION LOSSES OF NETWORKS

The insertion loss of the network becomes a very important factor to be considered, particularly as the power of the system becomes greater. Fig. 7-5 shows that an insertion loss as low as 1 db means that a 100-watt amplifier would deliver only 80 of its 100

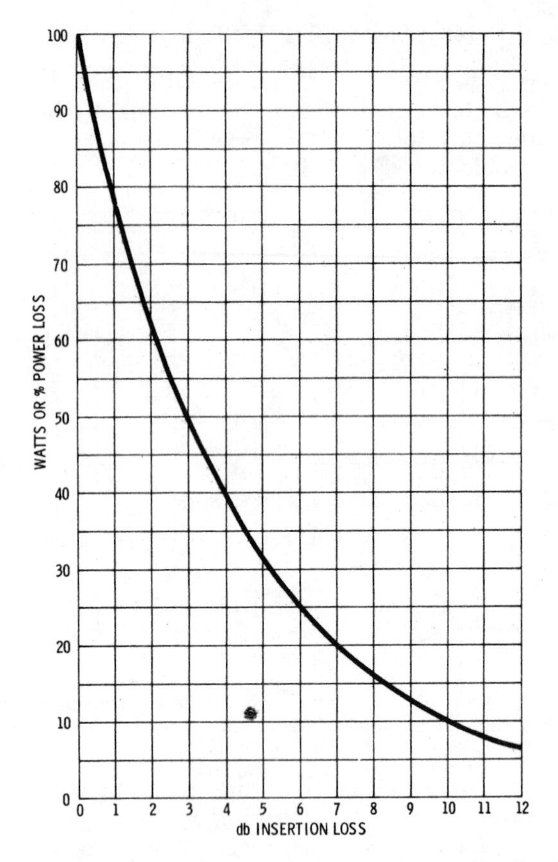

Fig. 7-5. Watts or percentage of power loss versus db of insertion loss.

watts to the drivers. An insertion loss of just 0.5 db would result in a loss of 11 watts of power at the 100-watt level. These losses could also be considered as percentages. Thus a 1-db insertion loss is a 20-percent loss of power; and a 0.5-db loss is an 11-percent loss of power.

CROSSOVER RATES

The steepness of the crossover "slopes" has varied over the years from 6 db per octave to 24 db per octave, with the rate of 12 db per octave becoming the most universal choice for optimum performance. A 6-db rate is too slow to protect the high-frequency driver at the lower frequency. At a constant power, the diaphragm excursion of the driver will increase by a factor of four, without a network, as the frequency is halved. As illustrated in Fig. 7-1, the power requirements rise steeply near the crossover point so that the high-frequency driver might be asked to produce appreciable acoustical power if only a rate of 6 db per octave were used. A rate of 12 db per octave ensures a rapid attenuation of the high-frequency unit as the electrical power increases in this critical area. Networks of 18 and 24 db per octave are so steep as to introduce transient distortion problems in the woofer and the high-frequency driver.

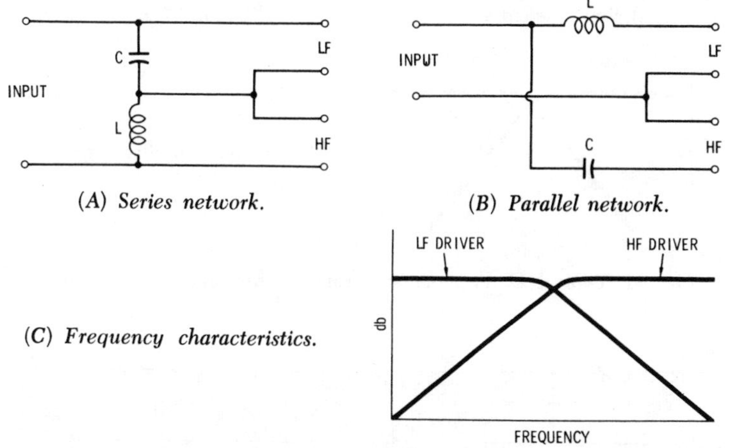

(A) Series network.

(B) Parallel network.

(C) Frequency characteristics.

Fig. 7-6. Typical 6 db-per-octave crossover network configurations.

ELECTRONIC CROSSOVERS

Electronic crossovers, where the woofer and the high-frequency driver are each driven by a separate amplifier, with the crossover network appearing at the input of the amplifier, create a dangerous situation for the drivers. If any low-frequency pulse occurs in the high-frequency amplifier, the high-frequency driver has no protection from it and will burn out. The amplifiers themselves basically must be wide-range units, even if only used to reproduce high frequencies. Wide frequency response is necessary to minimize phase shift in the desired bandwidth. Because of these problems in electronic crossover networks—the requirement of two amplifiers and the lack of protection to drivers—the passive network placed between the amplifier and the speaker completely dominates the field.

CHOKES

For the home constructor of crossover networks, air-core coils are the best to use, due to their simplicity of construction. The com-

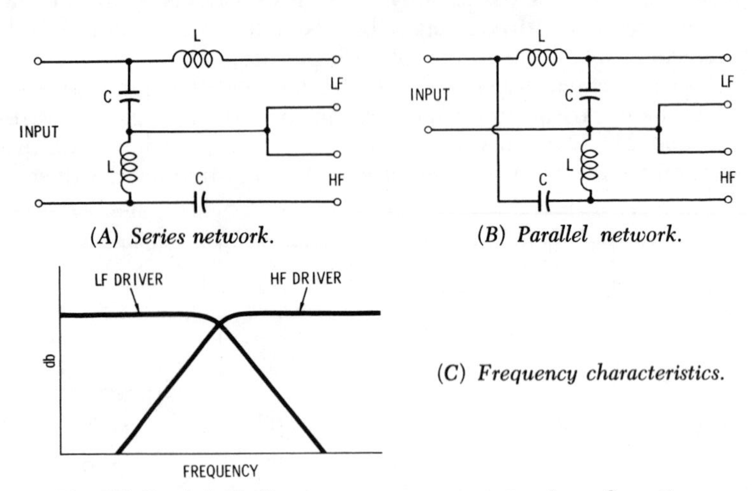

Fig. 7-7. Typical 12 db-per-octave crossover network configurations.

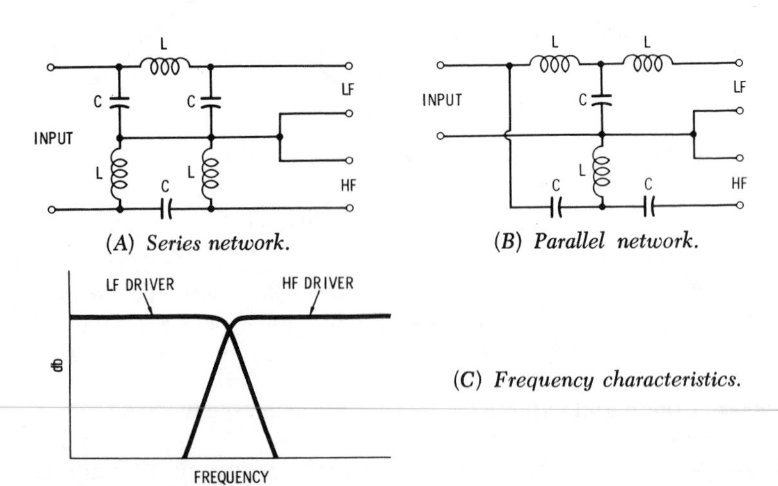

Fig. 7-8. Typical 18 db-per-octave crossover network configurations.

mercial designer of speaker systems has tended to use the iron-core coil to an increasing degree in recent years. The main reason for this is the ability, by very careful control of the coil-core design, to make an iron-core coil having less resistive loss than an air-core coil. If the core reaches saturation due to power overload, it becomes nonlinear and distortion is generated. Such saturation occurs at around 20,000 gausses. When iron-core coils are used commercially, they are usually designed for twice the expected maximum power input by use of a gap within the magnetic circuit.

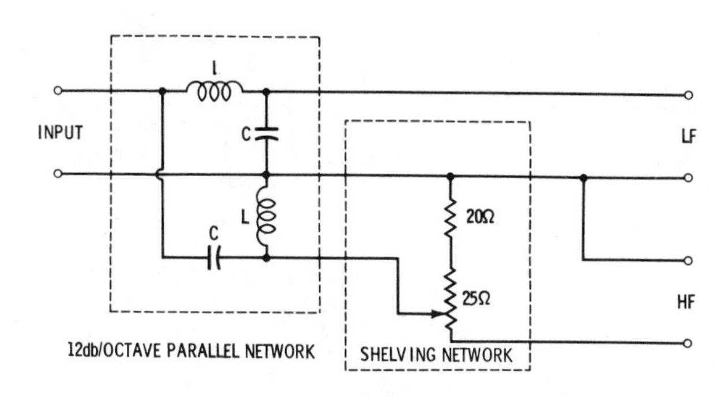

(A) Circuit diagram.

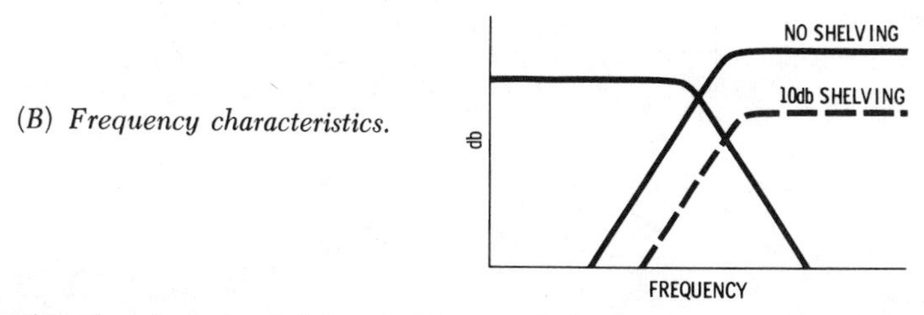

(B) Frequency characteristics.

Fig. 7-9. The effect of a shelving circuit on a 12 db-per-octave crossover network.

The iron core allows the same inductance to be achieved as in the air core but with considerably less wire and, consequently, less resistance. Networks using iron-core coils show insertion losses as low a 0.1 db compared to 1 to 2 db for air-core coil networks.

NETWORK CONFIGURATIONS

Of the various network configurations used there are constant-K networks of the series and parallel types, and the constant-resistance m-derived networks of both series and parallel types. Figs. 7-6, 7-7, and 7-8 show 6, 12, and 18 db-per-octave series and parallel crossover networks of the constant-K type.

EQUALIZERS AND SHELVING

Equalizers can be included in the crossover network if required. They are not really part of the network but can be combined with the network circuit.

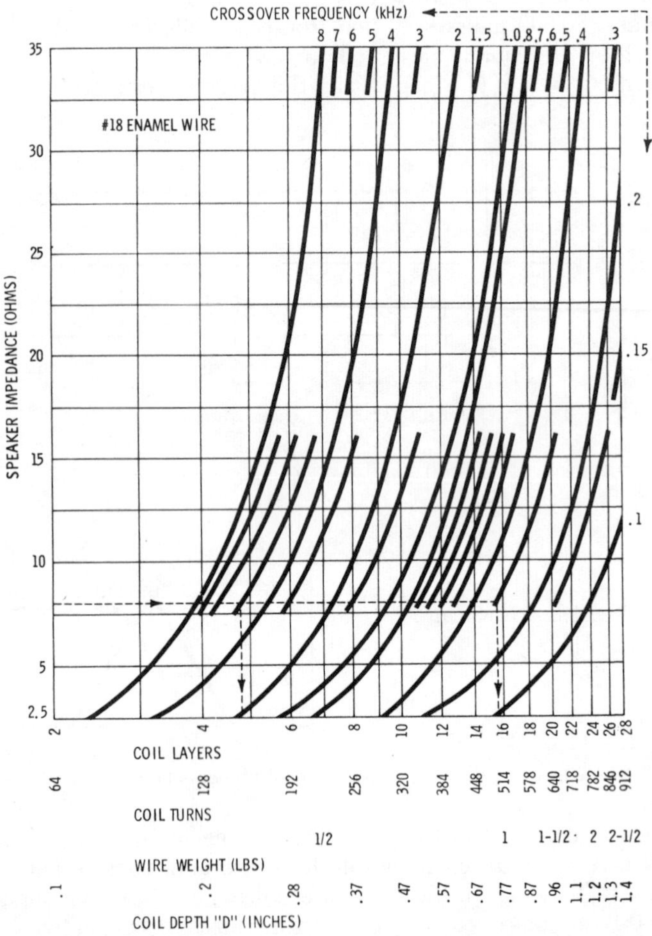

Fig. 7-10. This chart contains all the necessary winding data for constructing crossover-network coils. To use the chart, find speaker impedance on the vertical scale to the left. Move horizontally to the right until the curve for the desired crossover frequency is intercepted, then move down to the bottom horizontal line which gives winding data for a 6 db/octave coil (see dotted line example). For a 12 db/octave coil, multiply the speaker impedance by 1.41 and go through the same process. See Fig. 7-11 for coil dimensions that this winding data applies to.

Fig. 7-11. Physical dimensions of coil form for use with Figs. 7-10 and 7-12.

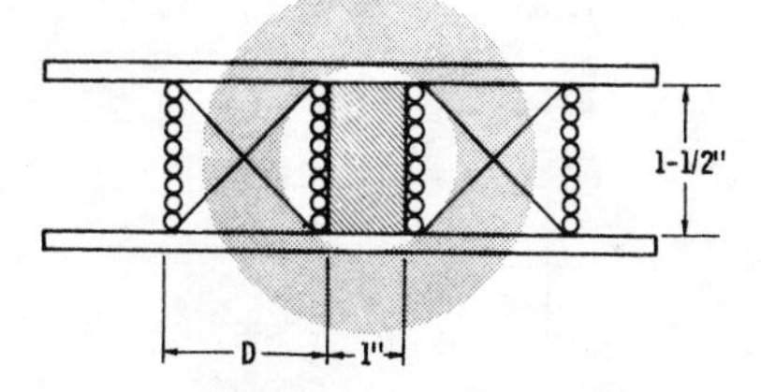

Since the low-frequency sections of most speaker systems have lower acoustical output than the high-frequency sections, some means must be provided to attenuate the high-frequency section. This is called *shelving*. Fig. 7-9 illustrates a shelving circuit added to a 12 db-per-octave crossover network.

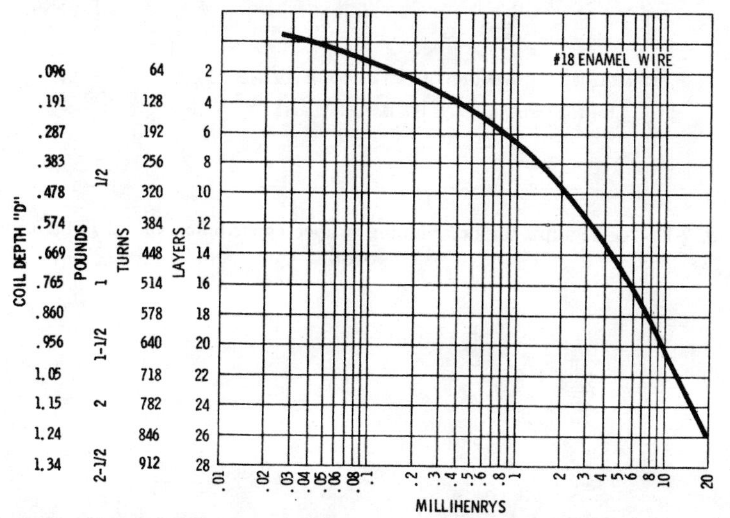

Fig. 7-12. This chart is for determining the actual inductance of coils wound from the data in Fig. 7-10 on the form shown in Fig. 7-11.

Shelving is also useful to reduce unfavorable interaction of the speaker with the acoustics of the listening room.

Figs. 7-10 through 7-12 show the construction details for coils of a given frequency and cutoff rate. Fig. 7-13 enables one to make the calculation of the capacitances required.

SUMMARY

With the completion of your crossover network, all that remains is to properly assemble all the necessary components, test them, and enjoy their successful performance.

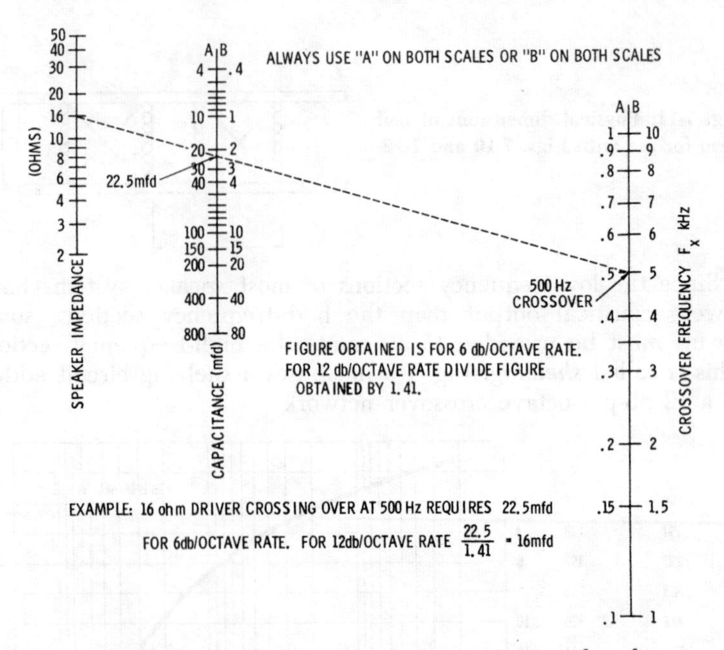

Fig. 7-13. Nomograph for determining proper capacitance values for crossover networks.

8

Construction and Testing Techniques

Once the driver or drivers, crossover network, and type of enclosure have been chosen, the careful physical integration of the components is the final step before testing and full enjoyment.

Each construction drawing included in this book outlines in detail the necessary bracing, gluing, fastening, and placement of absorbent materials, etc., to make an enclosure rigid and nonresonant. It is important to follow these plans carefully. They may also be used as the starting point in the construction of an enclosure of your own design.

The tools required are simple, and in most cases, familiar ones. They are:

1. A portable jig saw.
2. A portable rotary saw.
3. A quality portable hand drill, preferably ⅜ inch.
4. A Yankee-type ratchet screwdriver. (Many, many screws will be driven during the course of the enclosure construction.)
5. Hammer.
6. Clamps.
7. A selection of regular screwdrivers.
8. A selection of nutdrivers.

9. A soldering gun or iron.
10. Necessary measuring tools—rule, square, compass, etc.

RIGIDITY

Easily the most important consideration in the construction of a speaker enclosure is rigidity. A vibration of the enclosure means that useful energy is being lost. One way to ensure rigidity is to keep the individual panels small in size. Another way is to keep them in different sizes and different shapes, if possible. Enclosures are best when they do not have parallel walls. Triangular shapes such as used in corner-type cabinets are ideal. Where large, unbroken panels are used, they must be thick and cross-braced. It is important that the enclosure be airtight at all points other than the deliberately designed openings. Avoid creating any restrictions, cavities, or openings that are not called for in the plans. These can create resonance effects or organ-pipe effects.

In the process of maintaining airtightness, all joints should be treated with glue and then pulled tightly together with many wood screws. A screw every five inches is a reasonable estimate of the number of screws that should be used. The portable power drill should be used to start all holes so the screws do not split the wood.

Where glue is not used (on removable panels), even more screws should be used (one screw every three inches is about right). To ensure a really secure fit, thin rubber sheeting or felt can be used to form a gasket around the edges.

While the outer or decorative shell can be made of fine hardwood or other attractive material, the speaker enclosure should be made of plywood. The thinnest material that should ever be considered is ⅝ inch, and then only in corner-type cabinets where the natural triangular shapes aid in providing a rigid, strong structure. Most often, ¾ inch is the thickness used. Where extreme rigidity is needed, two sheets of ¾-inch plywood glued and screwed together result in weight and strength.

Use of metal for low-frequency enclosures is not desirable (unless a ¾-inch panel of lead is employed!). Normally, metal sheets have a tendency to "ring" at certain resonant frequencies.

Brick, stone, cement, etc., make suitable rigid materials, but are, of course, more difficult to work with and restrict mobility. Some constructors in years gone by have made enclosures with double walls and filled the space between with sand. All of these ideas are intriguing, but as has been steadily pointed out in this book, good engineering results in a balanced approach to the problem, and ¾-inch plywood represents, if properly handled, the very best results for the investment involved.

ABSORBENT MATERIALS

Naturally, great care should be taken to keep the drivers free of dirt, dust, and other construction debris during the mounting period. Many constructors shellac the inside of the enclosure after first vacuum cleaning the interior, to ensure smoothness and cleanliness of the surfaces.

Of the materials that can be used to provide absorption inside the enclosure (where called for) *Fiberglas* is the most universally used by manufacturers. It can be cut to any desired size and is very convenient to use. The *Fiberglas* should be tacked or stapled to either one of any two parallel surfaces to prevent standing waves within the enclosure. Typically, the bottom, back, and one side would have the material applied to it.

Corner enclosures, not having parallel walls, require absorbent material only at the bottom surface.

Small, compact, bookcase-size enclosures can be completely filled with absorbent material. This effectively increases the apparent volume of the box by a factor of 1.4, as discussed in Chapter 3.

DRIVER MOUNTING

Do not use screws to attach the drivers to the mounting panel. It is best to use ½-20 "T"-nuts so that the drivers can be bolted to the panel. This allows removal of the driver without the danger of the mounting holes becoming worn and undesirably loose. By doing this, the driver or drivers can be rotated at regular intervals (about once a year), which prevents excessive cone sag particularly among heavy woofers.

The same mounting system can be utilized for attaching the removable panel for driver access, whether it be the back panel or the side panel.

One caution: Large woofers have both large magnets and large paper-pulp cones. Many an unwary enclosure builder has had the unhappy experience of having his screwdriver or wrench deflected into the speaker cone by the strong field of the magnet while in the process of bolting the woofer to the mounting panel. Care should be taken to guide the screwdriver or wrench with both hands until it is safely coupled to the bolt head. It is also wise to remove wrist watches, etc., that might be adversely affected by the strong magnetic field of the woofer magnet. (Recently a shipment of high-frequency drivers was shipped by an air freighter from the West Coast to Chicago. The resultant magnetic field pulled the aircraft's compasses out of line, causing an unscheduled landing. The cargo had to be split up into several shipments.)

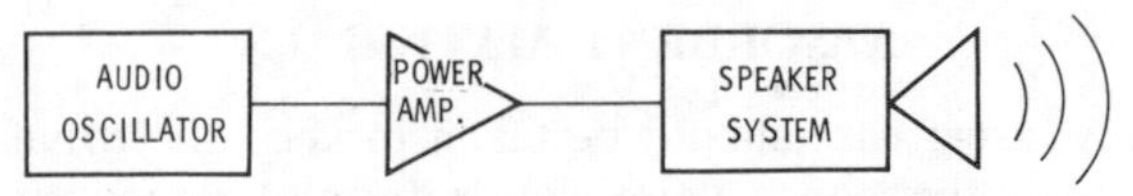

Fig. 8-1. Equipment setup to check for buzzes, rattles, and other noises.

Never use standard electrical plugs for speaker interconnections, due to the danger of accidental connection to a power circuit, which, of course, will burn out the voice coil.

GRILL CLOTH

Paint the speaker panel the same color as the speaker cone, usually black, so that there is uniformity of color under the grill cloth; otherwise, the speaker cone will show through. Touch up screw heads with the same paint.

Speakers work best without grill cloth; however, if grill cloth is used, it should be of the specially designed plastic-type cloth sold at any local audio-components shop. If one wishes to use a regular

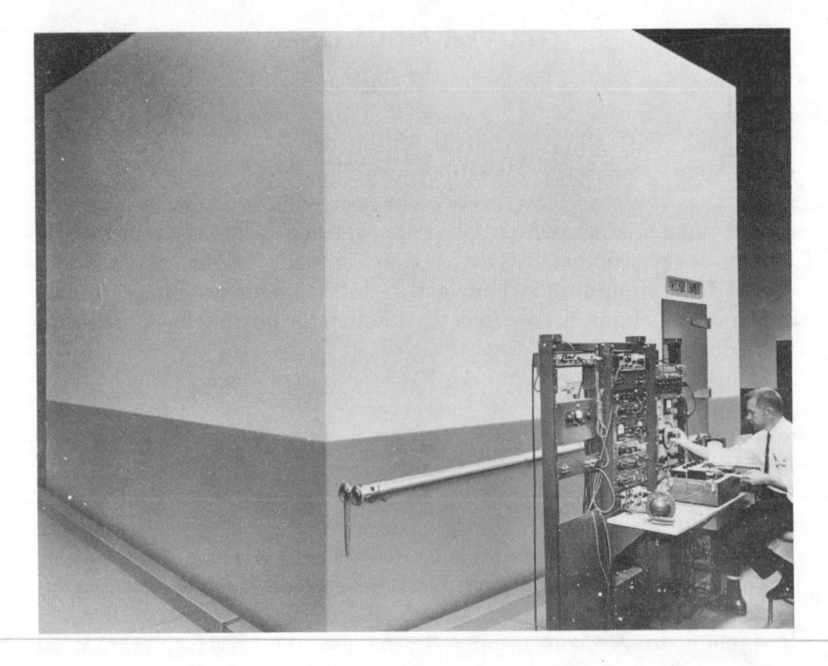

Fig. 8-2. An anechoic test chamber illustrating the large physical size required to make the chamber usable at low frequencies. These rooms are placed in sunken pits and are isolated from all building vibrations.

fabric, the plastic cloth can be used as a guide for proper density of weave. Care should be taken to stretch the grill cloth tightly and evenly across the speaker baffle so that it does not sag, wrinkle, or "drum." Grill cloth should be spaced away from the baffle surface to prevent it from slapping against the panel during the low-frequency excursions of the woofer.

Fig. 8-3. An interior view of the chamber shown in Fig. 8-2. Note the large *Fiberglas* wedges that help make the room "echo free."

TESTING

Once the enclosure is constructed and the drivers are mounted, the final remaining step is to adjust and test the system as a whole.

The following instruments are of inestimable value in adjusting, correcting, and testing the speaker system:

1. An audio oscillator of reasonably low distortion which allows continuous adjustment of its frequency.
2. A reliable a-c voltmeter preferably of the vacuum-tube type.
3. A good-quality oscilloscope intended for measurements at audio frequencies.
4. A calibrated microphone. (May be of the simple ceramic type.)

The first test to be undertaken is for buzzes and rattles (see Fig. 8-1). The only equipment required is an audio oscillator and the power amplifier that will be used in the system. A tone recording that sweeps very slowly up or down in frequency can also be used.

The system should be tuned to a medium volume at 1000 Hz, then the oscillator should be slowly swept down in frequency. Listen

Fig. 8-4. The equipment used in conjunction with the anechoic test chamber is costly, complex, and very precise.

carefully for buzzes, rattles, hisses, and any sounds other than the pure tone desired. Care must be exercised to distinguish between sympathetic resonances in the room (window panes, vases on shelves, lath in walls, etc.) and undesired vibrations of the speaker enclosure itself. When possible it is best to conduct such tests outdoors away from structures susceptible to such resonances. Buzzes (cone cry) usually are due to the speaker becoming overloaded. Rattles can be traced to loose grill cloth, loose panels in the enclosures, or the driver cone touching the mounting panel during its travel.

Hisses come from air leaks, particularly in the small, "shoe-box" size infinite baffles.

If a panel has not been properly fastened, or is in need of additional cross-bracing, pushing heavily against it with the hands will cause a change in the buzz or rattle. Speaker systems should be capable of being placed solidly in their location where they are to be used, and be free of extraneous vibrations. It is not always feasible to quiet all the sympathetic resonances in a room, but a smooth speaker system will tend to minimize such effects.

Figs. 8-2, 8-3, and 8-4 show the complexity of the testing facilities employed by a manufacturer in evaluating the products they design and produce.

For those who have access to similar laboratory-type equipment the following equipment setup diagrams can serve as a guide in making useful "living room" evaluations.

The use of pink-noise and constant-percentage bandwidth filters provides a source that has its energy distributed in a manner similar to that of musical program material.* Pink noise decreases at a rate of 3 db per octave as the frequency increases. Since a constant-percentage bandwidth wave-analyzer typically has a bandwidth which increases at a rate that causes it to read 3 db per octave higher for each octave as the frequency increases, on a signal of uniform energy distribution, the use of this type of analyzer with

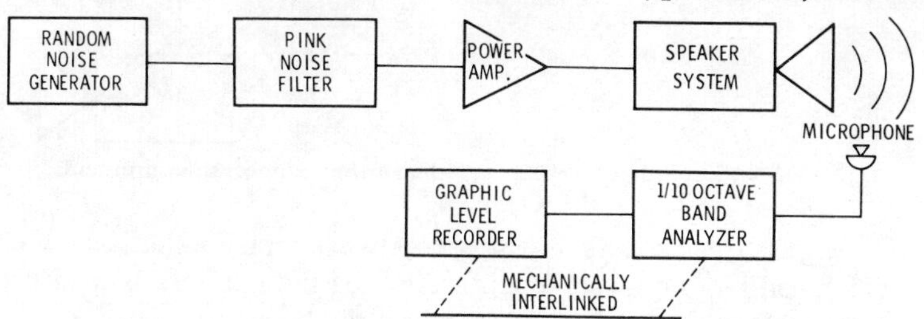

Fig. 8-5. Equipment setup for frequency-response measurements.

* Don Davis, *Acoustical Tests and Measurements* (Indianapolis: Howard W. Sams & Co., Inc., 1965) pp. 39-42.

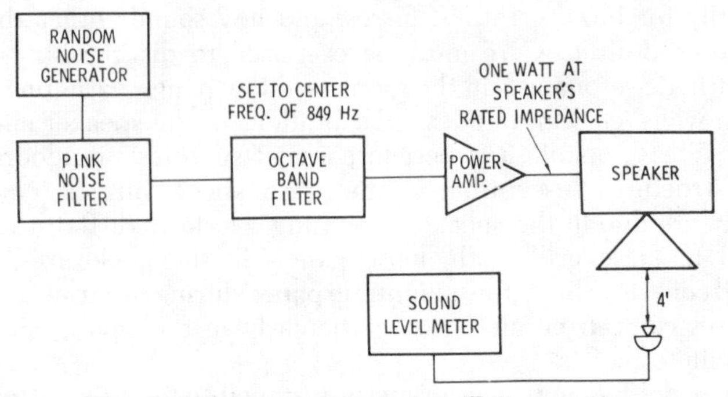

Fig. 8-6. Equipment setup for efficiency rating.

a pink-noise source plots on a chart as a "flat" line. By placing a microphone in the normal listening position in the room, frequency response curves can be run which show the interaction of the speaker system with the room acoustics at that particular location. (See Fig. 8-5.) In some very sophisticated systems, extensive broad-band equalization is done to provide a flat acoustical response at the listening position.

The useful efficiency of a speaker can be determined by the setup shown in Fig. 8-6. To measure the impedance of a speaker as a single figure given in ohms, the setup in Fig. 8-7 can be used. The variable resistor is adjusted until the a-c vtvm reads one-half of the voltage obtained at position A when switched to position B. The number of ohms the variable resistor reads when the voltage at B equals one-half the voltage at A is the impedance of the speaker system.

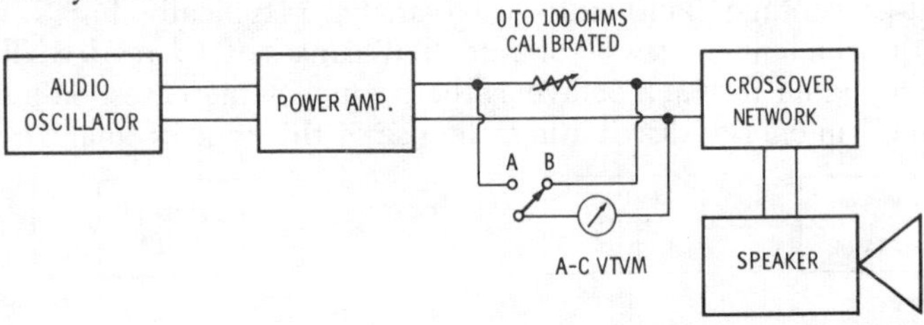

Fig. 8-7. Equipment setup for speaker-impedance measurement.

To measure a speaker crossover network, the method shown in Fig. 8-8 can be employed. Each section of the network is measured in turn. It is also interesting to measure the same network when the drivers themselves are used in place of R.

To measure the resonance of a speaker in open air, use the equipment shown in Fig. 8-9. Have the driver held up in the air free of reflecting surfaces. As the oscillator is swept down in frequency there will be an increased reading in the a-c vtvm as the point of the cone resonance is passed. (See Chapter 4 for details.)

To measure the distribution of sound in a listening room, the equipment setup shown in Fig. 8-10 should be used. The pickup

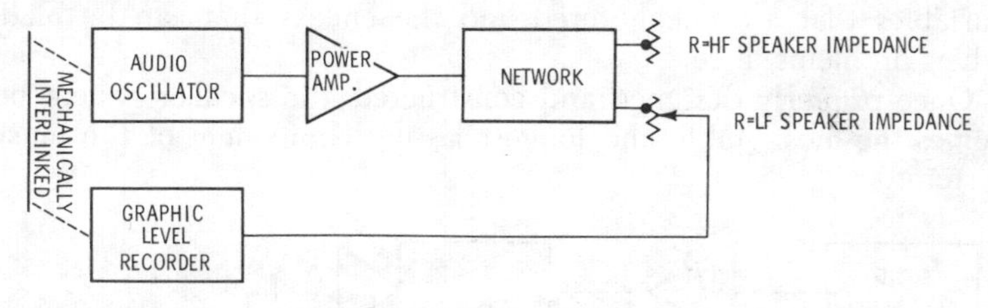

Fig. 8-8. Equipment setup for network measurement.

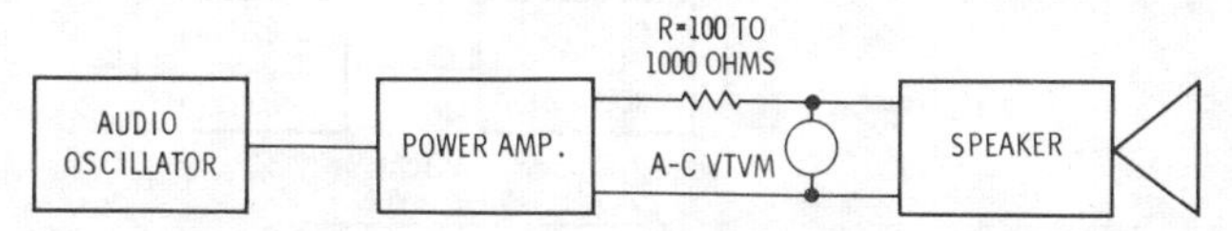

Fig. 8-9. Equipment setup for cone-resonance measurement.

microphone can be carried across the room slowly while the level recorder is allowed to run on a time-base chart. The microphone travel rate across the room should match some given number of intervals on the chart, and these intervals will then serve as a distance measurement.

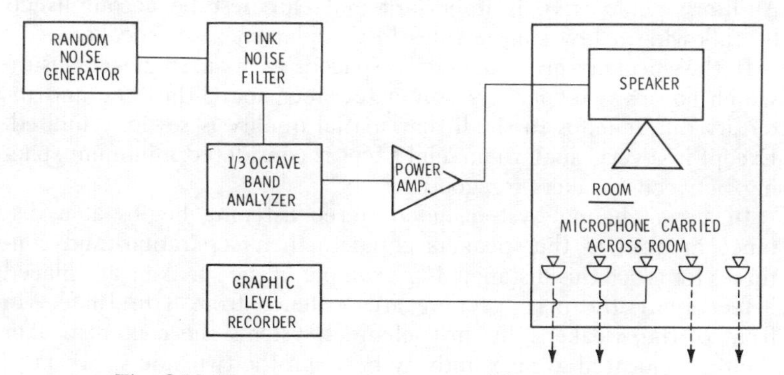

Fig. 8-10. Equipment setup for sound-distribution test.

Distortion can be measured if the level of the output from the speaker system exceeds the noise level (see Fig. 8-11) inherent in the listening room by at least 45-db spl. (In order to allow 1-percent recordings, a 40-to-1 db ratio is required—40 db = 100 to 1. The remaining 5 db is to ensure a safe margin.) Then, by sweeping the wave analyzer past a fixed signal from the oscillator, the harmonics, if present, will trace themselves on the level recorder.*

While these tests may never be performed, especially if one chooses to build from already tested plans rather than starting a design project, a knowledge of them allows some insight into the variables that are encountered and the checks that can be made when problems arise.

Once properly designed and constructed, the speaker system becomes the most stable and longest lasting component of the music system.

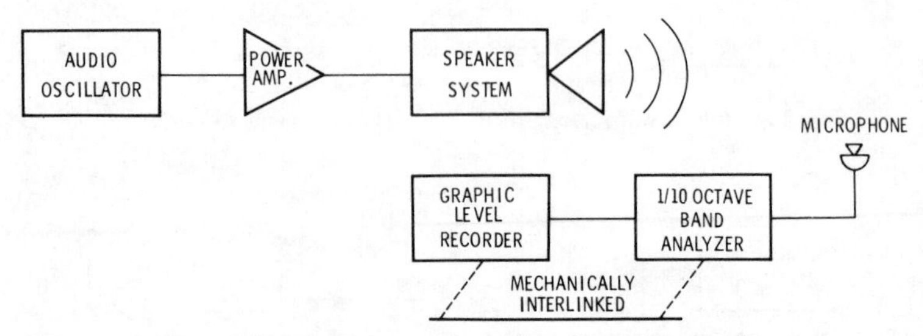

Fig. 8-11. Equipment setup for distortion checks.

STEREO SPEAKER PLACEMENT

There has never been any problem in the proper positioning of speakers when they are used for monophonic reproduction, but care should be exercised in speaker placement for stereo reproduction. Auditory perspective is important and can best be accomplished by following a few simple rules in the placement of speakers.

If the speakers are too closely spaced, as in a single enclosure which houses two speakers only a few feet apart, the time and intensity difference is so small that spatial quality is severely limited. Except in a very small room, eight feet is considered minimum spacing between speakers for good stereo.

In a two-channel system, good stereo listening begins at a distance in front of the speakers equal to their separation, and continues for twice this distance. For example, if the speakers are placed 8 feet apart, the best listening area extends from 8 to 16 feet in front of the speakers. In three-channel systems the center speaker should be located as near midway between the two side speakers as possible and ideally in line. The use of the third speaker allows wider spacing of the side speakers and a correspondingly broader listening area.

* For further testing information see *Acoustical Tests and Measurements* by Don Davis.

High-quality stereo home music components have provisions for connection of a center, or third-channel, speaker, or provide a voltage source capable of driving a third-channel amplifier.

In operation, the center speaker reproduces the signal existing in both side speakers. The sound that originated from the center of the recording stage, as in a recording of a soloist, will actually radiate from all three speakers but will appear to be confined to the center channel. With two speakers not too widely separated this effect is preserved without the center speaker; but the third speaker will permit a much wider separation of speakers and spread of sound source without an undesirable "hole in the middle" effect. It also allows corner placement of the two side speakers.

A large spread between speakers is desirable if the listening area can be moved back proportionately. Listening too close to two widely separated speakers creates a "hole in the center" which gives the impression of two distinctly separate sound sources rather than the desired broad front of sound. In a three-speaker system, the hole in the middle is much less likely to occur.

PHASING STEREO SPEAKER INSTALLATIONS

The relative phasing of the right- and left-hand speaker units in a stereo music system is essential, so that sounds meant to come from the center appear to emanate from a point midway between the two speakers. Many elaborate methods for determining the correct phase are available but by using a constant-amplitude, frequency record, available from any high-fidelity record dealer, it becomes a simple matter. The 100-Hz frequency band is recommended for this purpose.

Maintain consistent polarity in wiring your speakers, carefully following the instructions furnished with the speaker.

1. Listen to your system from a monophonic source while playing a 100-Hz tone and stand directly between the right- and left-hand speakers, about eight feet in front.
2. Reverse the polarity of either the left or right speaker, and if the volume "goes down," your speakers are out of phase; if the volume "goes up," they are in phase. They should be left connected in the loud position.

SUMMARY

Now you have finished your project. If you have designed well, constructed ruggedly, and tested and adjusted with care, just relax and enjoy years of good listening.

Index

D

E

F